INFORMATION TECHNOLOGY INVESTMENT

Decision-Making Methodology

INFORMATION TECHNOLOGY INVESTMENT

Decision-Making Methodology

Marc J. Schniederjans
University of Nebraska-Lincoln, USA

Jamie L. Hamaker
University of Nebraska-Lincoln, USA

Ashlyn M. Schniederjans
Johns Hopkins University, USA

 **World Scientific**

NEW JERSEY • LONDON • SINGAPORE • SHANGHAI • HONG KONG • TAIPEI • BANGALORE

Published by

World Scientific Publishing Co. Pte. Ltd.

5 Toh Tuck Link, Singapore 596224

USA office: Suite 202, 1060 Main Street, River Edge, NJ 07661

UK office: 57 Shelton Street, Covent Garden, London WC2H 9HE

British Library Cataloguing-in-Publication Data
A catalogue record for this book is available from the British Library.

INFORMATION TECHNOLOGY INVESTMENT
Decision-Making Methodology

ISBN 981-238-695-5

Printed in Singapore by World Scientific Printers (S) Pte Ltd

To my Myia, for your patience, understanding and support.

Preface

Periods of business activity are often marked by and referred to as "ages" in the historical development of the field of business. According to most business historians we have advanced from the "age of information" into the "age of knowledge". In both of these periods of time, information technology has been a determining factor in the survival and success of firms competing with one another. Those firms that know how to best invest in information technology have been and will continue to be the successors in this and future eras of business history.

Regardless of your position in an organization, investing in information technology may be the most important decision you will ever face in business. Unfortunately, investing in information technology is not as easy as common financial investment decisions. Careful consideration of financial and non-financial criteria may have to be included in the analysis to render an optimal solution. To make good decisions on information technology today requires the use of a variety of investment methodologies. These investment methodologies must be able to integrate the complexity of decision criteria in such a way that a decision choice is clear and clearly supported by the analysis. Today, just generating a decision is not enough. Information technology decisions must be supported by comprehensive inclusion of all relevant decision-making criteria.

The purpose of this textbook is to provide an in-depth treatment of a wide-variety of decision-making methodologies focused on the subject of investing in information technology. The methodological procedures as well as computer solutions to basic financial and advanced decision-making methodologies will be presented as tools for investing in information technology.

This textbook has been designed for an upper-level undergraduate course or a graduate business or engineering management course related to technology management for university students. Business faculty in areas of finance might also find this textbook useful for an applied investments course. Practitioners who work in information systems can also use this textbook if faced with a technology investment decision-making problem. Other groups of decision makers might include CEOs, vice presidents of information systems and finance, general managers, plant managers, supervisors, and industrial engineers. Other operations management and engineering faculty, trainers, and graduate students will also find this textbook present a useful variety of methodologies for managing and aiding information system investment decisions.

This textbook assumes that the reader has had some exposure to general mathematics and the terminology commonly found in business management. The basic technology/operations management or industrial management, and basic finance course that undergraduate students take for business degree programs, constitute sufficient prerequisite knowledge to satisfy the background to fully appreciate the content of this textbook. You do not have to be a technology manager or financial analyst to benefit from using this textbook. The terminology necessary to fully utilize this textbook is actually defined in the textbook. Also, throughout this textbook important terms are italicized and are usually followed by a definition. The location of the initial definitions can be found using the index at the end of the textbook.

The basic contents of this textbook are organized into twelve chapters consisting of four parts. In Part I, "Introduction to Information Technology Investment Decision-Making Methodology", three chapters are presented that help to define the basic subject and terminology used in the textbook, as well as briefly identifying the major topics that make up the rest of the textbook. In Part II, "Financial Information Technology Investment Methods", three chapters are devoted to explaining how basic financial methods are used and should be used in information technology decisions. In Part III, "Multi-Criteria Information Technology Decision-Making Methods", the three chapters presented describe and illustrate a more complex set of decision-making methodologies that can be used individually or in combination with other methods to render information technology investment decisions. Finally, in Part IV, "Other Information Technology Investment Methods", a series of three additional chapters present a variety of other commonly used investment methodologies reported in the information technology literature. Collectively these chapters provide a comprehensive treatment of commonly used and more recently applied methodologies for technology investment decision-making. This textbook ends with an epilogue chapter focused on the issue of not making the right decision and how the consequences might be avoided.

M. J. Schniederjans
J. Hamaker
A. M. Schniederjans

Contents

Lists of Tables and Figures

TABLES

Chapter 4

Chapter 5

Chapter 6

Chapter 7

Chapter 8

Chapter 9

FIGURES

Chapter 12

Epilogue

Part I

Introduction to Information Technology Investment Decision-Making Methodology

Part I

Introduction to Information Technology Investment Decision-Making Methodology

Chapter 1

Introduction to Information Technology Investment Decision-Making Methodology

Learning Objectives

After completing this chapter, you should be able to:

- Describe different types of IT investment decisions manager face.
- Briefly describe some of the methodologies that are used in IT investment decision-making.
- Explain why IT investment decision-making is important as a subject to study.
- Explain some of the limitations that should be considered when using IT investment methodologies.
- Explain the role of IT investment decision-making within organizational planning.

Introduction

The *productivity paradox* refers to the absence of a positive relationship between spending on *information technology* or IT and its resulting contribution to productivity or profitability (Lucas, 1999). Robert

Solow, the 1987 Noble Prize winning economist felt there was a singular absence of measured productivity from the use of computers when looking for it at the industry or economy level of analysis. Other researchers seeking to find a connection between capital investments in IT and productivity at the company or business firm level of analysis have been equally surprised to confirm the lack of a relationship between investment on IT and firm performance (Brynjolfsson, 1993; Landauer, 1995; Qing and Plant, 2001). However, several other researchers have found that there is a positive relationship between IT investments and firm productivity and performance (Bhatt, 2000; Dewan and Min, 1997; Stratopoulos and Dehning, 2000; Swierczek and Shrestha, 2003).

The inconsistency in the research results mentioned above can be viewed as a metaphor on the subject of IT investment decision-making. That is, there are no single, simple methodologies that will give a consistent, reliable and optimal solution to mangers facing an IT investment decision. One type of investment methodology can suggest one alternative and another methodology a completely different alternative to an IT investment decision choice. To try to help in this very complex decision situation, the purpose of this book explores a series of methodologies that can be used individually or in concert to help aid in IT investment decision-making. We will try to explain where these decision methods can be used, in most cases their mathematical computational procedures, their informational value, and their limitations.

In the next few sections of this chapter, we will briefly introduce some of the types of IT investment decisions managers face to provide an orientation to better understand the problems IT manager's face. We follow that section with an overview of the various types of methodologies available to aid in making those decisions and a brief discussion on their limitations. We also add an explanation as to why this subject is important to learn. We then describe the relationship of IT investment decision-making within organization strategic, tactical and operational planning as a way to bridge the context of general management. Finally, we end this chapter with an overview of the book organization to provide a logical system of for learning this subject.

Types of IT Investment Decision-Making Problems

Just about everyone has had the sometimes-challenging decision situation of purchasing a *personal computer* (PC). Did you ever think about the criteria you used as criteria and measures in coming to a final selection? Let's start with this type of simple IT selection problem as a common beginning to view the complexity in the differing types of decision-making problems this book will examine.

Most people start a PC purchase selection decision with a cost factor as a primary selection criterion. This criterion is usually measured or scaled in dollars as listed in Table 1. Within the dollar range (or sometimes beyond it) there are many other factors as briefly listed in Table 1 that create the typical multi-criteria problem we face in a PC selection. Note that some of the criteria are measured in dollars, some are ranked, and some are just noted as being present or not. While it may be easy to choose between computers based on an objective criteria measure, such as the size of its memory (i.e., larger is usually viewed as a better deal), how can a dollar be compared with the rank of a brand name? Yet, this is what thousands of people do every day when they purchase a PC. Some of them make good decisions for themselves, and sometimes not so good a decision.

A decision process that requires a sequence of decisions can further complicate the simple PC selection problem. Suppose a secondary feature, like DVD-ROM in Table 1 could be considered as add-ons to those PC's that do not have that feature. This type of sequence of decisions is depicted in Figure 1. This would mean that that the DVD-ROM feature (its costs, it quality, etc.) would have to be considered before a final decision on a PC could be made. This creates what is called a sequential decision process where a series of decisions must be made in an ordered fashion to arrive at the primary PC selection decision. This sequence could have many levels of decisions, each with multiple and conflicting criteria.

Table 1. PC selection criteria and measures.

Factors (Selection criteria)	Measures	Scale used in measure
1. Cost	Dollars	Number of dollars
2. Primary features	RAM speed	Processing time in mega bytes
	Memory	Processing time in gaga bytes
	Operating system	Ranking of brand name
	3.5 Disk drive	Present or not
	CD read/write drive	Present or not
	Word processing software	Ranking of brand name
	Speaker system	Present or not
	PC quality	Rating by consumer groups
3. Secondary features	Video card	Present or not
	Microphone	Present or not
	DVD-ROM	Present of not

Now, let's continue to complicate this PC selection decision situation by suggesting that the new PC is to operate with another of existing PC that the user owns. Now issues of compatibility of the hardware and software, as well as user required retraining on the new PC features have to be considered in the primary decision on which PC to purchase.

The PC problem above is very simple relative to what manager's face when selecting IT for operating systems in business organizations. In addition to all the factors above, business firms have to be able to integrate their systems within their own firms network and with their external partners, like customers and suppliers via the Internet or other mainframe computer information systems inside and outside the firm. These factors could include those reported by Sarkis and Sundarraj (2002): intrafirm adaptability, interfirm adaptability, platform neutrality and interoperability, scalability (resizing capacity to meet changing needs), security, system reliability, ease of use, and customer support. One very challenging factor Sarkis and Sundarraj includes is that managers must seek to justify a decision by showing that the investment in IT returns some form of "perceived value" to the firm.

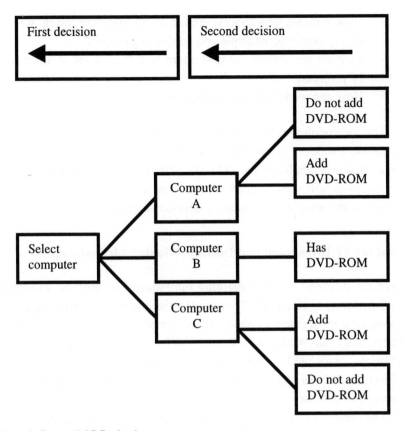

Figure 1. Sequential PC selection process.

With all that complexity, how can a decision on IT be made? It will not be easy, but this book helps to focus effort on identifying and using the right methodology to deal with the decision-making situation. The basic types of IT investment decision-making problems this book is focused on answering includes the following:

1. What are the most appropriate quantitative methods and techniques for the evaluation of IT?
2. What quantitative and qualitative measures can be used in the assessment and evaluate IT investments?

3. How can we objectively render an IT decision when we use highly complex, multiple and conflicting criteria?
4. How do we choose the best alternative from a set of alternative IT projects?
5. How can we justify our IT decisions?

The methodologies presented in this book can be used in a variety of situations, including decisions on technology, systems, software, and human resource applications. The illustrative applications presented in later chapters will seek to demonstrate some of the many possible areas that the methodologies can aid in IT decision-making.

What are IT Investment and Decision-Making Methodologies?

There are different ways of defining *information technology investment* or *IT investment*. Keen (1995) views IT investment as a term that applies to investing in equipment, applications, services and basic technologies. Others, such as Weill and Olson (1989) view IT investment as the expenses associated with acquiring computers, communications, software, networks and personnel to manage and operate a management information system.

The definition that we will use for purposes of this book includes all of the components that make up *management information systems* (MIS). All MIS's are a collection of four primary components: personnel, application software, system software, and hardware. As depicted in Figure 2, an MIS includes personnel who run and manage the information technology of the firm. The personnel might include users who must receive technology training (and therefore represent an IT investment), the technical personnel that perform the input/output functions of the system and run the operating computer systems, and their managers. Other components include the *application software* (i.e., programming languages, Assembly language, C++, etc.), and *system software* (i.e., operating systems, interpreters, compliers, utility programs to manage data, etc.). At the heart of all these personnel and software is the driving, interactive component of the IT hardware (i.e., computers, data storage disks and systems, communication systems, network systems, etc).

The relationship of resources allocated to the individual components as well as the collective system that makes up an MIS is the primary focus of this book. The definition of *IT investment*, therefore adopted for this book, can be defined as the investment decisions of allocating all types (i.e., human, monetary, physical) of resources to an MIS.

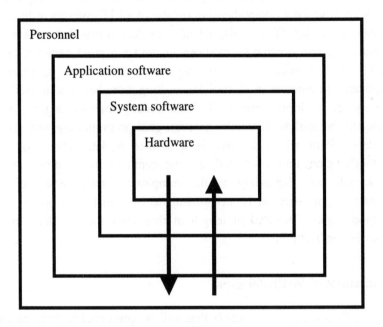

Figure 2. Management information system (MIS).

It is though the use of investment methodologies that we will be able to answer the questions this book posed in the last section. No book can possibly provide a comprehensive discussion of all the possible IT investment methodologies as there are differing opinions on what methodologies are relevant in IT and which are not. For example, Sylla and Wen (2002) suggest that cost-benefit analysis, return on investment, return on management, and information economics are the primary methods for evaluating the "tangible benefits" (i.e., benefits like profit or cost minimization).   They also recommend other methodologies for evaluating IT "intangible benefits" (i.e., benefits like customer satisfaction, improved employee motivation, etc.), such as multi-objective, multi-criteria methods, value analysis, critical success factors, methods for risk, real option, portfolio approach, and the Delphi method. Other studies on IT investment methodologies seek to just define a combination of quantitative and qualitative methods. For example Chan (2000) performed a comprehensive review of all IT investment literature in most of the top IT journals. Chan's conclusion was that it takes both quantitative and qualitative methods to render a good decision on IT. Sarkis and Sundarraj (2000) suggested that more sophisticated requirements inherent in IT decision-making were being addressed by a variety of multiple criteria decision-making methodologies, which included the analytic hierarchy process, goal programming, and scoring models (all of which we will be presented in later chapters).   The methodological topics selected and the depth of the treatment in this book will vary depending on the complexity and diversity of their potential application in IT investment decision-making. Some of the methodologies discussed in length in this book include those briefly described in Table 2.

Limitations of Methodologies

The methodologies and models that will be presented in later chapters are all based on the idea that we can identify and include all relevant IT decision-making factors. Unfortunately, the methodologies themselves limit what kinds of factors they can consider. For example, some investment models can use dollars as an input parameter, yet, as we have mentioned in the simple PC problem, it takes a combination of dollars of cost, a ranking of features, a possible scoring of quality, and so on, to render a more inclusive decision on IT investments.

In some situations not being able to include the right combination of decisions factors or criteria can limit what is considered in the final investment decision. Also, the factors that are included may have been measured incorrectly or contain bias in some way. This leads to the old modeling problem of "garbage in, garbage out."

Table 2. Select IT methodologies.

Schedule	Capacity
Analytical hierarchy process	Calculate the overall score of decision-makers' pair wise comparisons
Balanced scorecard	Evaluate investment from the user's, business value, efficiency, and innovation/learning perspectives
Critical success factors	Obtain, compare and rank factors critical to business success and based on these rankings, deduce investment priorities
Decision theory	Calculate the expected value of investing in alternative investments
Accounting rate of return	Compare the average after-tax profits with initial investment cost
Delphi method	Obtain consensus of experts' opinion concerning the best alternative investment
Satisfaction and priority surveys	Survey and compare user and MIS professionals' opinions on the effectiveness and importance of installed systems
Game theory	Calculate payoff of investment based on actions of the competition, mathematics and economic theory
Payback period	Calculate time required to recoup initial cost
Information economics	Calculate the overall value of an investment based on enhanced ROI, business domain, and technology domain criteria

Also, the time frame of comparing the costs of investing in IT and the eventual rewards or benefits of that investment do not always keep to a predicted time table. Usually IT investments involve upfront capital investment and the time period when the firm actually accrues the benefits of those investments may be very different, requiring a time adjustments component in the analysis. Also, financial-oriented methodologies tend to exclude most considerations of intangible benefits in preference to tangibles ones. Just measuring intangible benefits in such a way that they can be used in IT investment decision-making models is a very difficult task.

One of the primary areas that IT investment methodology is vulnerable is related to "risk". Investments in IT are subject to higher risks than other capital investments. This increase in risk is due in part to the fact that technology components are comparatively fragile, easily sabotaged by employees, and usually decentralized (e.g., data storage in one state and CPU in another), which leads to increased difficulties in IT design, development, management, and protection. Generally, there are two classes of IT risk:

1. *Physical risks*: The vulnerability of computer hardware, software, and data to theft, sabotage; software vulnerability to piracy and deletion; data security laps.
2. *Managerial risks*: Failure to achieve anticipated benefits or cost reductions: implementation failure to achieve a desired time frame; end-user resistance; inability of system to support organization or its growth over time; and incompatibility issues that later develop.

Unfortunately these risks and limitations are to a greater or lesser degree inherent in all capital investment situations, including IT investment decision-making. Fortunately, selecting the best model or models can minimize the risks. This is one reason why IT investment and decision-making methodologies are important to study.

Why Study IT Investment and Decision-Making Methodologies?

There are many reasons why IT investment and decision-making methodologies should be studied, but they collectively can be expressed as a means of achieving a *competitive advantage* (Laudon and Laudon, 2004, pp. 101-102; Turban *et al.*, 2001, pp. 447-449). Because management information systems are the core means of communication within firms and externally to customers and stakeholders, advancements in technology can quickly and efficiently give a competitive advantage of improved customer service. Their improvement can also allow a firm to more quickly seize business opportunities over their competitors. *Stakeholders* here can include more than stockholders or owners of a business, but also include the partnering companies that a firm counts on to help them perform their business functions. Examples of these partnering companies include transportation companies that ship and deliver a firm's goods. These partnering companies can also be consultants, subcontractors, drop-shippers, and all firms that support the operations and business functions of an organization. By improving the communication or ability to move data in a firm, you can multiply the efficiency and productivity of one firm over another, many fold. IT investments are the quintessential ingredient that can bring a quick and powerful improvement in communication and data movement, and thus bring a competitive advantage to a firm.

On the other hand, if a firm poorly invests in IT, that investment can become a competitive disadvantage needlessly increasing capital costs, increasing interest costs, delaying customer orders, disrupting communications within the firm and other stakeholders, and decreasing employee morale. These costs can be considerable. In a survey reported in *Computerworld* (1999) the time to implement an enterprise-wide IT system takes 23 months with an average cost of US $10.3 million dollars. And these are only the up-front costs that a firm will be out if the system fails.

One purpose of this book is to help insure that IT investment decisions achieve a competitive advantage and help avoid any of the competitive disadvantage situations. Firms can only hope to realize their goals and objectives in IT investments if they carefully make their decisions on the best possible information. The IT investment decision-making methodologies that are presented in this book are designed to provide additional information on which, at least, a better informed decision can be made.

Organizational Strategic Planning in IT Investment Decision-Making

It is important to know where IT investment and decision-making methodologies fits into the general planning framework of a firm. To understand their role lets begin with basic MIS hierarchical planning stages as presented in Figure 3. There are three basic stages of planning in all organizations, and in all functional areas, such as the functional area of MIS (Irani and Love, 2002; Laudon and Laudon, 2004, pp. 72-101). At the *strategic planning* stage senior managers are expected to be involved in developing specific systems to implement corporation-wide strategy, and also develop the strategies themselves (Adler, 2000). This planning might involve deciding on expanding IT resources to support an expanded supply-chain distribution system for the corporation. It might involve the weighing of the risks of those expansions and the need to justify them within the context of corporation mission or purpose statements. The outcome of this stage of planning is usually a general set of goals and objectives, as well as some priorities and very general longer-term time-tables for their accomplishment. Firm often confirm compliance of these goals and objectives with corporate governance mandates (O'Donnel, 2003). For most organizations the corporation-wide goals (like growth in sales or growth in facilities) are also broken down into how the functional areas can support them. For example, a growth in sales for the corporation might be supported by the functional area of MIS by developing a new e-commerce division strategy for added e-commerce sales.

At the *tactical planning* stage it is expected that middle-level managers will implement the goals and objectives defined at the prior strategy stage. The planning now becomes a matter of how to implement the stated goals and objectives. While the strategic plan might have a five year schedule, the tactical plan would break this down into smaller time periods, usually what must be done each year to accomplish the longer-term strategic goals. This planning also breaks the work down from one general set of strategic plans for all of the MIS functional area, to individual MIS departments or divisions. In this way the planning become more specific in time and units of effort to be performed. The tactical stage is chiefly focused on allocating resources to have the capacity achieve the desired work. An example would be determining the number of programmers needed each year for the next few years necessary to install a enterprise-wide computer network. Most importantly, it is at this stage of planning that the investment decisions on IT are made.

Finally, at the *operational planning* stage the more detailed, day-to-day work effort is planned and scheduled. An example here is a the monthly, weekly or even daily schedule of work load of each employee, in each skill grade. Where tactical planning would consider total employees in a department, operational planning is much more detailed and focused on unique individual skill requirements to accomplish the more general tactical goals and objectives in scheduling work on a daily basis.

The MIS hierarchical planning stages in Figure 3 are fairly broad and general. Lets now narrow this planning effort down to more specific MIS steps or tasks to better see where the role of IT investment decision-making is positioned in organizational planning. These steps can be broken into the nine steps in Figure 4 (Michaud and Theonig, 2003; Wheelen and Hunger, 2003; Kangas, 2003; Hill and Jones, 1992).

Step 1. External analysis of competition and threats: In this step an analysis of the firm's external environment is undertaken in order to determine the major threats and opportunities facing the organization. This would include an analysis of the general environment, consisting of technological factors (e.g., speed of change in some IT is greater than expected, may outdate current investments), political factors (e.g., competition has newer computer systems and is viewed as more up to date than our firm), economic factors (e.g., competition is spending more on IT), physical factors (e.g., do we have the space or capacity to make IT changes equal to the competition), and social factors (e.g., does our competition have better skilled people than we do). This analysis also includes risks that are posed by customer's expectations (e.g., customers expect our firm to be the most advanced in IT), suppliers (e.g., new ordering technology used by suppliers requires our firm to update to be competitive), competitors, and regulatory groups (e.g., changes in law mandate required investments in technology in order to comply with new regulations).

From this analysis we determine what opportunities we might have to beat our competition and areas where they pose a threat to our organization.

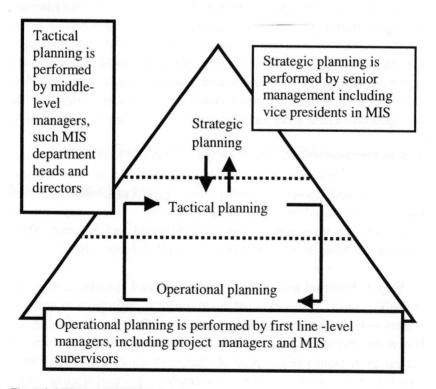

Figure 3. MIS hierarchical planning stages.

Step 2. Internal analysis of the firm's strengths and weaknesses: An analysis of the firm's internal resources is undertaken in order to determine the organization's major strengths and weaknesses. These strengths and weaknesses can stem from the firm's structure, culture, and functional area resources. A firm's strengths and weaknesses could revolve around factors such as:

1. Culture and how it promotes a high service level and employee loyalty.
2. Organizational structure and how it promotes flexibility and innovation.
3. Financial resources and how they give the firm the ability to obtain new equity and provide a steady cash flow.
4. Human resources and how they include quality managers as well as providing the firm with cost efficient labor, achieving a desirable absenteeism rate, and minimizing worker turnover.
5. Technical resources that promote high service level and employee efficiencies.
6. Physical resources that allow for flexible facility and equipment requirements and/or economies of scale.
7. Organizational resources that include an effective management information system, good coordination of functional departments throughout the organization, effective marketing, and/or a good public image.

Step 3. Overall corporate strategic planning: As previously stated, we seek here to achieve a corporation-wide policy that is consistent with the firm's corporation mission statement and general goals.

Step 4. MIS functional area strategic planning: We seek here to individualize the corporation-wide goals into the more narrow aspects related just to the functional area of MIS. This completes the strategic planning stage of the process and we move to the tactical steps next.

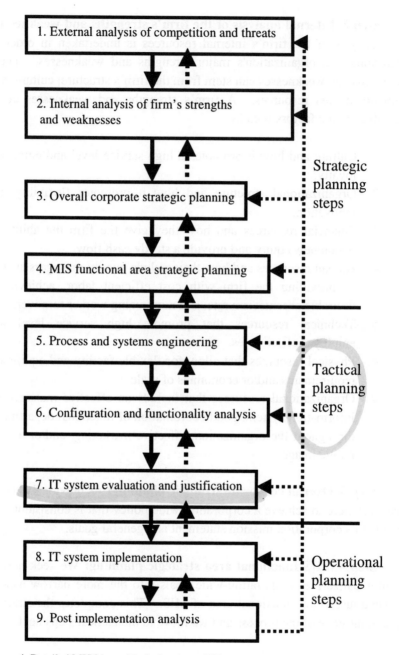

Figure 4. Detailed MIS hierarchical planning of IT systems.

Step 5. Process and systems engineering: This analysis involves a thorough development and determination of the inputs, outputs and business processes of the firm's systems. This includes collecting cost and benefit information. The idea is to provide a base-line in which to measure the future impacts of change brought about by changes we may make in IT. This step might involve the process of business process reengineering, where we look at current policies, practices and procedures in delivering products to all customers (external customers and internal users), and see if a revision in those policies, practices, and procedures might lead to an improvement in service or productivity.

Step 6. Configuration and functionality analysis: This analysis depends on what is being considered in the change process. Usually it involves exploring alternative IT configurations (e.g., alternative network configurations). These configurations are examined in terms of how well they function to serve areas of business operations, such as marketing, sales, manufacturing, finance, accounting, maintenance, engineering, and human resources. In this tactical step of the analysis a variety of quantitative and qualitative IT investment methodologies can be utilized. For example, multi-criteria methodologies like the *analytic hierarchy process* (AHP) can be used to rank differing configurations with relation to their ability to provide enhanced customer service. Thus AHP can help in the selection of the most ideal choice of configurations that will provide the best functionality and best customer service.

Step 7. IT evaluation and justification: Clearly this step in where all the IT investment methodologies are brought to bare on selecting and evaluating the best IT alternatives. This might include system-wide choices that are not accomplished in Step 6, or it might include very detailed individual component choices that make up a system. In this step we may find the sequential-type of decision-making problems complicating the decision process and multi-criteria compounding decision issues. Once the IT evaluation and justification is completed, we move to operationalize the decision in the next steps involving operational planning.

Step 8. System implementation: System implementation can be divided into four steps: acquisition and procurement, operational planning, implementation and installation, and finally integration. This is a very difficult step and often requires overcoming many difficulties because of differences in subsystems, platforms and interfaces. There are several strategies that can be used to accomplish this step and should be considered in the IT investment analysis as presented in Table 3.

Table 3. System implementation strategies.

Implementation strategies	Description	When used
1. Direct conversion	An existing system is removed totally and a new system installed	If there is only capacity or space allowed for one system to operate at a time or that the existing system is too costly, or dysfunctional to ongoing operations
2. Parallel conversion	An existing system and new system operate simultaneously until the new system is fully functional and the existing system can be discontinued	If the cost of shutting down the existing system in prohibitive
3. Phased conversion	New system is phased in as modules are systematically brought online	If the architecture of existing system will permit the gradual updating of new modules or the costs of a completely new system are beyond the resources of the firm
4. Pilot conversion	New system is fully implemented on a pilot basis in one part of the business operation	If the system has features that need to be examined in use, or the risk of converting entire system is too risky or expensive

Step 9. Post implementation analysis: This is a critical step that closes the loop of the IT planning process. While all the steps with the dashed lines indicate that feedback is possible to make revisions from the prior step, if necessary, this ninth step is a final form to check against the goals and objectives set at all the strategic, tactical and operational planning stages. It is a final check to make sure that the cost and benefits observed in Steps 6 and 7 and expected in the new system benefits are achieved in the end that was developed over all nine steps in this MIS planning process.

In summary, the use of the IT investment decision methodologies can support and aid in several steps in the overall MIS planning process, most notably, Steps 6, 7 and 9.

How This Book is Organized to Help You Learn

Structurally, the remaining chapters in this book have common educational pedagogy designed to aid the reader in understanding the text material each chapter seeks to present. In addition to the basic subject content, each chapter includes the following seven sections:

1. *Learning Objectives*: Readers should use these declarative sentences as a statement of what they can expect to find in the chapter and as a review tool after they have read the chapter to ensure they have attained the basic knowledge objectives of that chapter.
2. *Introduction*: Readers will find a helpful overview of the organization of the content of the chapter.
3. *Summary*: At the end of each chapter a brief summary of the chapter is presented in the first paragraph to remind readers of major points and on occasion discussion limitations of topics.
4. *Review Terms*: Throughout the book when new terms are introduced they are italicized and are restated here to remind readers of their importance. This listing also serves as a quick guide to abbreviations.

5. *Discussion Questions*: A set of discussion questions are presented as a means to stimulate ideas on content and further thinking.

6. *Concept Questions*: These questions can be used as assignments or a self-testing check to see if readers have learned the basic topics of the chapter.

7. *References*: All the references used for materials throughout the chapters are listed here.   Readers can use these reference citations to locate the publication and further their knowledge of specific content mentioned or referenced in the chapter.

8. *Problems*: In some chapters where methodology is quantitatively presented, a set of problems are presented for assignment purposes and also to help readers understand computational aspects of the methodologies, while expanding their understanding of how they can be applied in IT investment decision-making situations.

This book's chapters are organized into four parts. Part I, "Introduction to Information Technology Investment Decision-Making Methodology," consists of three chapters. In Part I, Chapter 1, "Introduction to IT Investment Decision-Making Methodology," a basic overview was presented of what this book is focused on and how it is related to the planning of management information systems. Creating a beginning foundation for what IT investment decision-making methodology involves, this first chapter's content is further refined in the following chapters. In Part I, Chapter 2, "Needs Analysis and Alternatives IT Investment Strategies," an examination of justifying IT investments is presented. This chapter identifies the need to explore alternatives IT investment strategies as a means to justify the IT investment plans necessary to run state of art IT-based firms. Issues such as outsourcing IT needs and other alternatives are explored. In Part I, Chapter 3, "Measuring IT Performance" the issues and problems of IT measurement are presented.   These issues include the economics, business performance, efficiency and effectiveness measures and each are examined to help establish a basis of consideration in the evaluation of IT investments.

In Part II, "Financial Information Technology Investment Methods," three chapters seek to present a variety of financial investment methodologies to aid in IT investment decisions. In Chapter 4, "Basic Financial Methods" classic financial-related methodologies such as breakeven analysis, payback period, and accounting rate of return methodologies are described and illustrated. In Chapter 5, "Other Financial Methodologies," a variety of differing classic financial methods are presented, including present value analysis, return on investment methodology, internal rate of return, and cost/revenue analysis. In Chapter 6, "Cost/Benefit Analysis," is presented.

In Part III, "Multi-Criteria Information Technology Decision-Making Methods," a variety of non-financial management science methodologies are presented in three chapters. These methodologies include: Chapter 7, "Critical Success Factors, Delphi Method and the Balanced Scorecard Method", Chapter 8, "Multi-Factor Scoring Methods and Analytic Hierarchy Process", and Chapter 9, "Decision Analysis and Multi-Objective Programming Methods." These chapters' contents are based on the most recent research on IT investment decision-making in their respective areas.

In Part IV, "Other Information Technology Investment Methods," we finish our presentation with three additional chapters describing a variety of combined financial and non-financial methodologies that are commonly used in IT investment decision-making. These methodologies include: Chapter 10, "Benchmarking Techniques and Game Theory", Chapter 11, "Investment Portfolio Methodologies", and in Chapter 12, "Value Analysis and Satisfaction/Priority Survey Methods."

We end our book with an Epilogue chapter, "The Costs of Not Making the Right IT Investment Decision," as a way of reminding readers that there are consequences of not exercising good judgment and good methodology in making IT investment decisions.

Summary

In this introductory chapter we have introduced a definition for IT investment decision-making and explained its relationship in the MIS planning process.   We have tried to explain the importance of this subject and briefly discussed its limitations in helping to aid in IT decisions.   We have described where in the multiple stages of MIS planning that the application of IT investment methodologies are applied and fits into the overall planning of an organization.

One of the main points this chapter makes is that IT investment decision-making is a difficult task. While a handful of methodologies are briefly mentioned in this chapter, they are but a small number that exist to support IT investment analysis and decision-making.   Some organizations actually avoid having to make IT investment decisions altogether by exploring alternatives to IT investment.   This type of avoidance behavior can occur as a result of what is called "needs analysis" and is the subject of the next chapter.

Review Terms

Analytic hierarchy process (AHP)

Application software

Competitive advantage

Direct conversion

Information technology (IT)

Information technology investment

Management information systems (MIS)

Managerial risks

Operational planning

Parallel conversion

Personal computer (PC)

Phased conversion

Physical risks

Productivity paradox

Sequential decision process

Stakeholders

Strategic planning

System software

Tactical planning

Pilot conversion

Discussion Questions

1. Why is the "productivity paradox" important in IT investment decision-making?
2. Why is there such diversity in the types of IT investment decision-making problems?

3. What is the relationship between the components of an MIS and the use of IT investment decision-making methodologies? That is, give examples of the MIS components that might require an investment.
4. Why is it important to consider the limitations of IT investment decision-making methodologies in an analysis?
5. Why is it important to see where IT investment decision-making fits into the overall planning of business organizations?

Concept Questions

1. How does "sequential decisions" add complexity to a particular decision situation?
2. What are four types of IT investment problems? Explain each.
3. How would you describe three IT investment decision-making methodologies?
4. What are the four components that make up an MIS? Are decisions on IT supported in all four areas?
5. What are the three stages of MIS hierarchical planning? How are they further broken down it nine different steps? Where does IT investment decision-making fit in to the hierarchical plan?

References

Adler, R.W., Strategic Investment Decision Appraisal Techniques: The Old and New," *Business Horizons*, November-December, 2000, pp. 15-22.

Bhatt, G.B., "Exploring the Relationship Between Information Technology, Infrastructure and Business Process Re-engineering," *Business Process Management Journal*, Vol. 6, No. 2, 2000, pp. 139-163.

Brynjolfsson, E., "The Productivity Paradox of Information Technology: Review and Assessment," *Communications of the ACM*, Vol. 36, No. 12, pp. 67-77.

Chan, Y.E., "IT Value: The Great Divide Between Qualitative and Quantitative and Individual and Organizational Measures," *Journal of Management Information Systems*, Vol. 16, No. 4, 2000, pp. 225-261.

Computerworld, "ERP Averages" *Computerworld*, April 5, 1999, p. 6.

Dewan, S. and Min, C., "The Substitution of IT for Other Factors of Production: A Firm Level Analysis, *Management Science*, Vol. 43, No. 12, 1997, pp. 1660-1675.

Hill, C.W., and Jones, G.R., *Strategic Management: An Integrated Approach*. Boston, MA: Houghton Mifflin, 1992.

Irani, Z. and Love, P.E.D., "Developing a Frame of Reference for Ex-ante IT/IS Investment Evaluation," *European Journal of Information Systems*, Vol. 11, 2002, pp. 74-82.

Kangas, K., *Business Strategies for Information Technology Management*, Hershey, PA: Idea Group Publishing, 2003.

Keen, P.G.W., *Every Manager's Guide To Information Technology: A Glossary of Key Terms and Concepts for Today's Leader*, 2nd ed., Boston, MA: Harvard Business School Press, 1995.

Landauer, T.K., *The Trouble with Computers*. Cambridge, MA: MIT Press, 1995.

Laudon, K.C. and Laudon, J.P., *Management Information Systems: Managing the Digital Firm*, 8th ed., Upper Saddle River, NJ: Prentice Hall, 2004.

Lucus, H.C., *Information Technology and the Productivity Paradox*, Oxford, UK: Oxford University Press, 1999.

Michaud, C. and Theonig, J.C., *Making Strategy and Organization Compatible*, New York, NY: Palgrave/Macmillan, 2003.

O'Donnell, A., "Lean & Mean," *Insurance & Technology*, Vol. 28, No. 7, 2003, p. 27.

Qing, H. and Plant, R., "An Empirical Study of the Casual Relationship Between IT Investment and Firm Performance," *Information Resources Management Journal*, Vol. 14, No. 3, 2001, pp. 15-26.

Sarkis, J. and Sundarraj, R.P., "Factors for Strategic Evaluation of Enterprise IT," *International Journal of Physical Distribution and Logistics Management*, Vol. 30, Nos. 3/4, 2000, pp. 196-220.

Stratopoulos, T. and Dehning, B., "Does Successful Investment in IT Solve the Productivity Paradox?" *Information and Management*, Vol. 38, No. 2, 2000, pp. 103-117.

Swierczek, F.W. and Shrestha, P.K., "Information Technology and Productivity: A Comparison of Japanese and Asia-Pacific Banks," *Journal of High Technology Management Research*, Vol. 14 No. 2, 2003, pp. 269-289.

Sylla, C. and Wen, H.J., "A Conceptual Framework for Evaluation of IT Investments," *International Journal of Technology Management*, Vol. 24, Nos. 2/3, 2002, pp. 236-261.

Weill, P. and Olson, M.H., "Managing Investment in Information Technology: Mini Case Examples and Implications," *MIS Quarterly*, Vol. 13, No. 1, 1989, pp. 3-17.

Wheelen, T. and Hunger, D., *Strategic Management and Business Policy*, Upper Saddle River, NJ: Prentice-Hall, 2003.

Chapter 2

Needs Analysis and Alternative Information Technology Investment Strategies

Learning Objectives

After completing this chapter, you should be able to:

- Describe the steps in the "IT project" planning.
- Describe a "needs analysis" and explain how it is used.
- Describe the types of costs that are considered in a needs analysis report.
- Describe the composition of an "IT project team" and their purpose.
- Explain how "reengineering" can be used in IT investment analysis.
- Define "IT outsourcing" and explain its role in IT investment analysis.
- List and explain strategies for IT outsourcing.

Introduction

In the previous chapter we outlined an overall management information systems (MIS) planning process that is made up of strategic, tactical, and operations steps (Hill and Jones, 1992; Michaud, and Theonig, 2003). We also indicated that this book is focused on the tactical steps (note Figure 1) of this broad MIS planning process (Wheelen and Hunger, 2003; Kangas, 2003). What this chapter seeks to do is to refine into greater detail the tactical planning steps and explain how to begin their implementation.

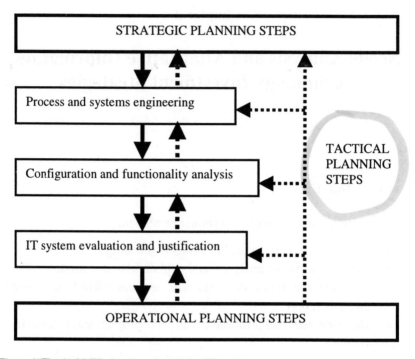

Figure 1. Tactical MIS planning process for IT projects.

No firm uses exactly the same set of steps to undertake their acquisitions of IT. What is being proposed in this book is a general framework that offers a logical ordering of tasks or steps that generally will be performed to accomplish "IT projects" of any shape or size. We are using the term *IT projects* here to mean any IT acquisition, including individual pieces of hardware, software, human resources, or an MIS in total. The fact is, when any component or change in a MIS system is suggested, it requires a rather complete analysis of the "system" for possible impacts. So a fairly similar MIS analysis or planning process is undertaken to plan these changes and their potential impacts. Why we can't just make a change and see what happens is because of the potential for a "ripple effect." A *ripple effect* can occur when you change an IT component of a system, and that change causes other components (e.g., say older versions of integrated software applications) to discontinue providing the same functionality or capacities to the user prior to the change in the one component. A component change usually requires the entire MIS to be examined for the impact of the change in the one component, as well as its relationship with the system as a whole. For example, some IT projects are mandated by government regulations and others are simply a process of upgrading existing software. These types of decisions don't seem to require any really serious thought, yet they do. If a firm is to acquire a new version of one software application, that firm must still decide on when the implementation must be undertaken, which technologies to use (if more than one manufacturer sells them), and how the change might impact existing older software systems that must be integrated with it. What if the other versions don't integrate with the new version of this one software application? The result may be a ripple effect on other existing technology, which in turn might require further changes and acquisitions to make the entire system work together as it did prior to the single software change. Those acquisitions could create further integration problems with other software existing in the system, and so on. With proper IT project planning, the ripple effect can be anticipated and minimized.

The tasks that make up the three tactical steps in Figure 1 can be further refined or divided into the five steps in Figure 2. The focus of this chapter is on the first four steps of Figure 2 and the methodologies we will present in the remaining chapters of this book are applicable to the last step of the MIS tactical planning process steps. We will begin with Step 1 and introduce a conceptual approach to the MIS planning process for IT projects called, "needs analysis."

What is Needs Analysis?

While most of the steps in Figure 1 are self-explanatory, we begin with a conceptual approach to determining what a firm's IT needs are as a first step (Kendall *et al.*, 2000; Shriver and Wold, 1990; Sylla and Wen, 2002). A *needs analysis* consists of determining what technology, software, human resources or a complete management information systems (MIS) are necessary to achieve an organization's stated strategic goals and objectives. While the needs analysis is an important first step

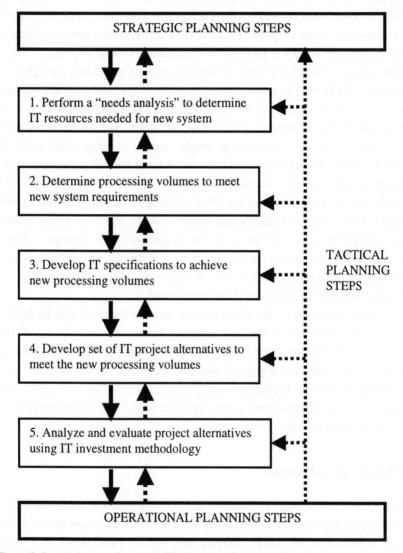

Figure 2. Detailed steps of tactical MIS planning process for IT projects.

in the process of making an IT investment decision, it is actually a tactical step in the overall MIS planning process presented in the previous chapter (note in Chapter 1, Tactical steps in Figure 4). Since the strategic planning precedes this tactical effort, the strategic decision that some kind of change is necessary has already been made and it now falls to IT managers to develop tactical plans and strategies to implement that change (whatever it may be). So, we start here with a set of MIS strategic goals and objectives to achieve. To develop these strategic goals and objectives into tactical plans managers must know two things that the needs analysis seeks to provide: (1) a current status or capacity of the firm's IT, and (2) future needs that define what new or additional requirements are necessary to help the firm achieve their strategic goals.

The current status, called a *needs report*, seeks to document the present MIS system or state of IT capacities. This information is usually found in a firm's policies or procedures manuals, computer application reference manuals, flowcharts of input/output forms, interviews with MIS personnel, and system checklists. It can sometimes be achieved by simply asking questions of the MIS personnel who operate the system where change is being planned. Questions like "What hardware is currently being used in your department?" and "What hardware is not currently being used in your department?" can obtain some of the desired information for this part of the needs report. Since the systems exists and has been providing information processing services, people who have worked with the system, such as MIS managers and supervisors, staff engineers, and in some cases cleric help, can all be sources of this descriptive information on what the existing investment in IT can do in terms of transaction processing.

To obtain a detailed listing of future needs requires considerable information collection efforts. Questionnaires can be used here again to ask questions like: "What hardware or additional equipment do you need in your department to implement the new procedures?" Since the new system might require new IT investments, the knowledge of current IT staff and their suggestions are almost always augmented by a host of additional outside experts. Experts on technology trends and equipments, consulting industrial engineers knowledgeable on IT and user-interface issues, software and hardware vendors, and IT service vendors who offer service contracts are often a part of the team of experts used to advise on IT issues. Indeed, an *IT project team* is often formed from a group of these experts in combination with MIS managers and supervisors to study and undertake IT projects. It is necessary to have an assemblage of these types of specialized skills when considering the complexities in business processes and systems, as well as alternative network configurations and functionality issues of IT. Move over, there is also estimation effort required to include projections of service and product processing capacities to be able to insure the new IT will be of use now and will be able to meet future needs. This might necessitate consulting forecasting experts and even futurists familiar with trends in IT. All of these efforts are necessary to complete Steps 1 to 3 in Figure 2 of the tactical MIS planning process for IT projects. Indeed, vendors, who may eventually bid on work that will be contracted outside the firm, also are a common source of information in establishing cost and time estimates in this estimation process.

In addition to establishing IT needs, the IT project team is usually charged with the responsibility of establishing the potential scope and size of the proposed project. This portion of the needs report would define what management is willing to spend on the IT project, what personnel will be involved, if outside consultants might be used to provide services or advice, deadlines for project completion, what costs and benefits might be realized, and how the costs and benefits will be measured. It will also include, as best as can be estimated, what software applications will be changed/removed, the potential impact on business transactions, what new networking requirements are necessary, requirements on physical hardware and the needs for communication within and outside the organization. This report will also provide some assessment of the complexity of the IT project so later costs, time and other resource comparisons can be fairly made between alternative technologies.

Determining IT Processing Capacities and Specifications

Cost assessment is an important part for all three of the first steps in IT project planning. Cost considerations should include items such as those presented in Table 1 (Kendall *et al.,* 2002). These types of information are eventually used in a document called a *request for proposal* (RFP). The RFP can be the outcomes of Steps 2 and 3 in the IT selection process in Figure 2 because of their detail.

Table 1. Costs considerations in the needs analysis report.

Type of cost	Description
Purchasing, renting, leasing	Investments in hardware can be purchased outright, rented, or leased.
Software	While bundling makes software application difficult to compare, the costs of the basic purchase decision can and must be compared to make a decision. Other support services (e.g., discounts on upgrades, diagnostic support, etc.) must also be factored into the decision process.
Personnel	Some systems require additional personnel and others only require a reassignment of existing staff. This would also include management costs
Hardware maintenance	Monthly or yearly fees that are required as a part of a purchase agreement on the IT.
Warranty	Some warrantees are included in the purchase price of IT and some require additional fees for maintenance service or replacement. Other extended service contracts might also be necessary to permit an equal comparison with differing vendor purchase offers.
Communication	The costs of communicating data from a main facility to intra-organization and extra-organization facilities. Cost estimates might be based on size of data facilities, type of line (e.g., dial-up verses dedicated), line speed (i.e., baud rate), and line quality.

This document can include requirements on each of the IT project alternatives covering factors such as proposed hardware, systems and application software, service contracts, personnel training, conversion requirements, future expansion requirements, implementation schedules, use of new or refurbished equipment, and physical facility locations. How we can determine what needs to be changed and explore business processes, procedures and IT can be accomplished using a variety of engineering and management study analyzes. Some analyzes are based on forecasting transaction processing capacities into the future and others can be focused on very specific issues, such as network configurations. One of the most common general engineering approaches used to deal with all aspects of determining IT processing capabilities and specifications is called "reengineering management."

A *reengineering program* is a process of drastically or radically changing people, processes, and the organization itself, through the use of technologies and methodologies, to achieve organizational objectives such as improved efficiency, quality, and competitiveness (Chase *et al.* 2004, pp. 338-341; Schniederjans and Cao). *Reengineering management* also called *business process reengineering* (BPR) are the activities that are involved in the managing of the restructuring of processes in an effort to improve efficiency.

Many organizations have sought to use a reengineering management program as a means to rapidly enhance service quality in an effort to develop a competitive advantage. It is also an ideal means of identifying where change in systems should take place and a means of identifying alternatives to improve systems. Research has shown that IT investments can play an important role in strategic planning for organizations during a BPR program (Bhatt, 2000; Bhatt and Stump, 2001).

A procedure for conducting a reengineering program is presented in Figure 3 (Hammer 1997; White 1996). As shown in Figure 3, Step 1, the organization starts the program after having been given a set of strategic objectives to achieve. In identifying the goals and particularly the specifications for change a number of sources of information may be called for in Step 2. For example, "benchmarking" (see Chapter 10) might be a source of goals and specific specifications to improve *business processes* (i.e., a very large set of interrelated tasks, departments, and organizations).

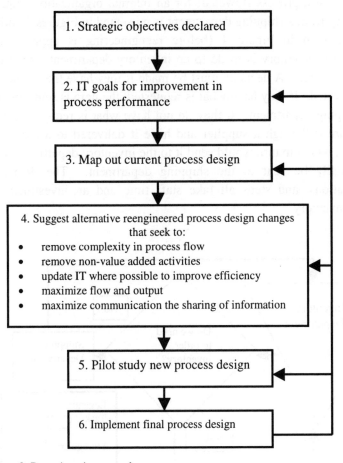

Figure 3. Reengineering procedure.

Other technical methodologies can also be employed, such as "gap analysis" (we will discuss "gap analysis" in Chapter 7). In Step 3 the current business process must be mapped or described. One of the primary methodologies to do this is with a *process map*, also called a *process flowchart* which is a simple graphic aid used to define each of the elements that make up a process. They allow managers to see how inventory flows through a process so opportunities for improvement can be identified and implemented during the reengineering of the process. An example of a process flowchart for an internal organization request for inventory from a shipping department is presented in Figures 4 and 5. As can be seen in Figure 4 (before reengineering is applied) the requisition for inventory is made to an inventory department where the inventory is stored. A decision must be made if they have the inventory that is requested. If they have what is requested, the item is sent directly to the shipping department. If they do not have what is requested, they must acquire it through a supplier and have it delivered to a receiving department, which in turn would send it to the inventory department so it can eventually be sent to the shipping department. The decision, communications, and steps all take staff time and an investment of capital in inventory.

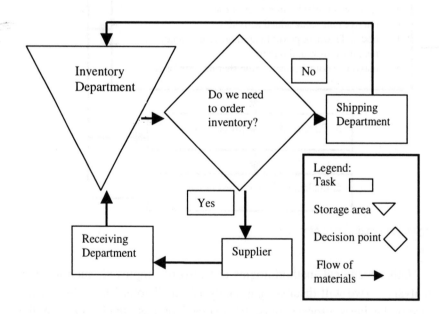

Figure 4. A process flowchart (before reengineering is applied).

Moving to Step 4 in the reengineering procedure in Figure 3 we would explore some possible changes in the inventory requisition process to reduce the complexity of the process and reduce time and wasted effort. While many strategies and guidelines can be employed to redesign changes in the process, one way to reduce complexity and inventory is to employ a principle of eliminating inventory by "outsourcing" the inventory function and allowing the supplier to maintain and supply all inventory requirements (we will discuss "outsourcing" later in this chapter). This would result in eliminating the inventory department and the capital investment in inventory. We might also require our supplier to deliver the inventory directly to the point of need within the shipping facility. This would eliminate the need for a receiving department. As a result of these changes in the role of the supplier, the resulting reengineered inventory requisition process would look like the process flowchart in Figure 5. This would constitute the temporary design that would, in Step 5 (note Figure 3) of the reengineering procedure, have to be pilot tested to see if it would be feasible and work to achieve the desired goals. This pilot testing would fall into Steps 4 and 5 (note Figure 2) of the MIS planning process. Step 6 in Figure 3, falls into the operational planning (note Figure 2) steps of the MIS planning process.

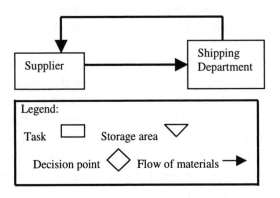

Figure 5. A process flowchart (after reengineering is applied).

In summary, the first three steps of MIS planning process for IT project planning (note Figure 2) helps the IT project team or the MIS manager know the current status of their IT resource capacities to meet the stated strategic MIS goals and objectives, and develop a listing of IT specifications that define new levels of resource capacities necessary to fully achieve those goals and objectives. In Step 4 (note Figure 2) of the MIS planning process for IT projects we need to determine a set of alternatives that will meet the desire specifications.

Alternative IT Investment Strategies

In Step 4 of the MIS planning process for IT projects two parts of analysis must take place to enumerate the IT alternatives that we may be facing in this investment situation. The first part is logically based on the information and specifications obtained in the previous analysis. That is, we set out a listing of IT alternatives that can meet the specifications in the RFP and meet other requirements identified in a reengineering program. This part of the planning effort is usually performed with the help of technology vendors who submit bids defining costs, services, quality, etc. This planning effort can also be accomplished by hiring experts or consultants whose knowledge of IT places them in a better position to know what IT is best to purchase based on factors of obsolesents, cost, quality, etc. As listing in Table 2, strategies for change can range from continuing without change to a complete change over of systems.

The other part of Step 4 (note Figure 2) is to explore the possibility of alternatives "to" an IT investment. That is, explore the options that might save a company from having to acquire some or all of what has been stated in the needs analysis as IT resources or make any change to their existing systems that is viewed as necessary to achieve their goals. This is an important part of the entire IT project planning process because it can help to justify the need for an investment if these alternatives to IT investment are themselves infeasible.

Table 2. Alternative IT investment strategies.

Alternative	Description
Continue using system without change	Continuing with the present system with little or no change assumes the present in-house processing capacity is adequate for present and future needs. This is the least costly and demanding alternative. If, on the other hand, the present system is not adequate in functionality, processing capacity, or it is no longer supported by its vendor, then this strategy may not be appropriate.
Reorganize system	Firms have reaped economies of scale by reorganizing their IT technology into "service centers." A *service center* typically is a physical location where data processing and IT services are provided at a main facility to branch locations geographically dispersed. By concentrating IT in one location it is assumed they can be better utilized with less management and staff. As a result it is expected that fewer staff and other IT resources are needed to accomplish the save level of work. The result is increased productivity, better IT utilization, and lower costs of processing transactions. If, on the other hand, no real economies of scale are possible, then this strategy may not be appropriate.
Upgrade existing system	Minor or major system changes with a series of upgrades to modules can permit a less costly and less potential disruption to the system as a whole, particularly with regard to the integration of modules and the retaining of personnel who will be familiar with the existing systems. If, on the other hand, the upgrade does not provide the desired functionality or limits processing capabilities in the future, particularly where the business processing environment is highly unstable and requires quick adjustments to market conditions, this strategy may not be appropriate.
Replace system	Replacing entire modules or enterprise-wide systems can be very costly and time consuming. This strategy can result is a quick turnaround and substantial improvement in functionality and processing capacities. If, on the other hand, the downtime due to learning the new system or implementing such a drastically new system is greater than the firm can afford, this strategy may not be appropriate.

Indeed, one of the first questions asked by bankers or finance managers in a firm, when asked to provide the funds for IT investments, is, "Do you really need the IT to accomplish your goals?" This part of the analysis helps to answer that question and may in some cases, save further unneeded analysis if the answer to the question is in fact "no". One alternative that can save further IT project analysis while still providing a means to achieve new IT processing capabilities is through a process called "IT outsourcing."

What is IT Outsourcing?

In situations where a firm does not want to tie up some its capital resources in MIS's it can hire the MIS tasks to be performed by outside companies or vendors. This process of contracting out the MIS tasks is called *IT outsourcing* (Laudon and Laudon, 2004, p. 399). The MIS tasks include more than staff members (like temporary help), but can include all technology, human resources, and management. Indeed, as can be seen in Table 3, the outsourcing strategies can include the entire MIS department if necessary.

Outsourcing IT is a very common MIS management planning decision. In a survey of 150 IT executives, 93 percent reported that they had outsourced some IT and business process functions (Lackow, 2001). Greaver (1999, p. 14) reported that outsourcing IT had increased 20 percent yearly through the 1990's. Lackow (2001) also reported the areas commonly outsourced included user support, disaster recovery, software development, software maintenance, support services, Internet services, and business processes. This survey revealed the main reasons for outsourcing where, among other things, cost savings, improved service, and access to outside expertise. It was also observed in the survey that the business executives polled had either fully achieved (55 percent of the respondents) or partially achieved (39 percent of the respondents) their objectives, while only 6 percent felt they had not achieved their objectives using outsourcing as a strategy. It also appears that outsourcing from nations like the US going to continue well into the future (Thibodeau, 2003). This study reported that 80 percent of the US businesses where planning some top-level negotiations on offshore outsourcing in 2004 and 40 percent of the companies have completed some kind of pilot program so they will be using near-shore or offshore services. The main reason reported was cost differences between the US and other importing nations. In 2003 for example U.S. software companies were charging $80 to $120 per hour for programming work, while the fee for offshore providers is about $40.

Table 3. Outsourcing alternative strategies.

Outsourcing strategy	Description
Subcontracting, limited work assignments	Short-term, over flow work beyond existing capacity is assigned to the subcontractors or vendors. This is just a temporary assignment in much the same way that temporary staffers are hired to fill in for summer vacation assignment of full-time staffers. This strategy is ideal with security issues or when cost prohibits more inclusion from a subcontractor.
Subcontracting, project assignments	Whole IT projects are assigned to subcontractors or vendors. These assignments would entail a complete project where the management of the project would be delegated to the subcontractor and not under the control of the MIS staff of the hiring firm. This strategy is ideal when a company has unique skill or technology requirements too expensive for them to maintain but affordable for contractors to offer their clients.
Total outsource assignment	Where part (i.e., staff, IT, facilities, etc.) or all of the entire MIS function is subcontracted out to a subcontractor or vendor. Here a company may lease all their IT from a subcontractor but run the equipment with their own staff. This strategy is ideal when a company may have a market that requires constant changes in IT or can not afford to tie up capital in IT.

Outsourcing has resulted in a number of benefits for users (Chiesa *et al.*, 2000; Currey, 1995; Lackow, 2001). These include:

1. *Technology improvements*: Allows the hiring firm to avail itself with the latest in technology.
2. *Financial gains*: Reduces IT payroll and moves the costs from fixed to variable. Permits the salvage sale of existing equipment and recouping of some of those capital expenses.
3. *Reduces budgets*: Bidding by subcontractors on a yearly basis permits ever increased pressure on subcontractors and vendors to reduce costs, thereby reducing budgets costs.
4. *Productivity improvement*: The latest in technology usually brings with it improved productivity and efficiency that the firm could not afford if it had to make the entire capital investment.
5. *Tax benefits*: Firms can deduct outsourcing costs from current income, in contrast to depreciating computer hardware purchases over a number of years.
6. *Enhances core business activities*: Allows firm to concentrate its limited staff on work that is at their core while permitting other MIS tasks to be performed by vendor.
7. *Facilities management*: Acts to insulate the firm from unexpected shifts in processing transactions, which would other wise require the firm to incorrectly invest longer-term capital in IT to handle short-term demand requirements.
8. *Management planning*: It is easier for most firms to outsource unexpected fluctuations of processing transactions than to plan and adjust in-house MIS resources to handle them.

There also have been a number of problems reported by managers with the outsourcing strategy that has resulted in a discontinuing of this strategy (Currey, 1995; Earl, 1999; Lackow, 2001; "Outsourcing Information Technology: Cutting Costs Or Cutting Your Throat?", 2003; Sislian and Satir, 2000). These can include:

1. Failure to achieve client needs;
2. Poor service quality; and
3. Failure to meet timely objectives.

There are also a number of barriers that inhibit or limit the use of IT outsourcing (Lackow, 2001; Currey, 1995), which include:

1. Questionable past and observed performance;
2. A lack of experience with outsourcing in general;
3. Organizational resistance to change; and
4. Inadequate preparation and planning.

Care must be taken in choosing outsourcing partners because of the relationship that can develop between firm and the outsourcing partner. Some of the guidelines in implementing an outsourcing strategy that are offered in the literature include (Currey, 1995; Turban *et al.* 2001, pp. 497-499):

1. Develop specific and complete service agreements and measure to monitor their compliance.
2. Include cash penalties for non-compliance.
3. Establish a trial period in which to evaluate the outsourcer.
4. Measure all IT activities during the trial period and report performance to outsourcer.
5. Include clauses in agreements that permit the firm to alter business operations if need be.
6. The firm should closely manage the inevitable cost escalation in contracted services.
7. The firm should closely measure and manage service quality.
8. The firm should avoid becoming too dependent on the outsourcer.

An IT outsourcing strategy

One IT outsourcing strategy that has emerged to take advantage of most of the benefits and avoiding some of the possible problems, is called "selective IT outsourcing" (Chiesa *et al.*, 2000; Lacity *et al.*, 1999). *Selective IT sourcing* can be a combination of all three of the strategies in Table 3 but with emphasis on only outsourcing tasks, projects and departments where the outsourcing will make a substantial improvement over existing IT facilities. The idea is to let subcontractor who can do tasks better than the firm, handle those task, while retaining those tasks that the firm does better than others. This permits a reallocation of resources away from tasks they don't do as well as a subcontractor, and reallocating those IT resources to those tasks they do better than others. For example, a firm might have a competitive advantage over competitors in marketing, but is not able to do as well as competitors in, say, processing payments. So, the firm would outsource the payment processing transactions to a vendor and take those investments in IT that supported the payment processing and reallocate them to other MIS tasks.

The advantages of selective IT sourcing include (Greaver, 1999; Lacity *et al.*, 1999):

1. Reduced personnel;
2. Reduced politics from internal MIS functional department staff;
3. Increased flexibility in organizational structure;
4. Enhanced ability to make rapid changes to meet changes in the business environment;
5. Reduction in capital expenditures;
6. Ability to provide new services with little investment;
7. Overcoming a lack of in-house expertise;
8. Accelerate reengineering benefits;
9. Gain access to world-class resources and capabilities;
10. Reduce and control costs of IT and
11. Share risks with outsourcing partners on IT investments.

Most importantly, selective IT sourcing improves the allocation of resources within the firm, resulting in better usage of financial capital, and a general improvement in areas where the firm previously was not as successful. The results of using this strategy will undoubtedly cause an improvement in the way the business operates, thus helping the firm become more competitive.

Final Comment on IT Outsourcing

It would be great to just outsource all IT effort and let someone else deal with those problems. Unfortunately, as noted above, we live in a real world where such idealism can get us into trouble. We know that improvements and advances in IT requires firms to constantly revise and change their investments in IT at a faster rate than any other type of capital investment. Indeed, it seems like newer technologies and software are made available on an almost daily basis, causing the need for constant reassessments in order to remain competitive. This may explain why prior research has reported such large increases in IT budgets for most firms every year.

What if IT outsourcing turns out to be too expensive or just not the strategy a firm wants to use to plan there MIS needs? That outcome can and surely does happen, but the analysis performed on IT outsourcing helps to justify why it is necessary to go on with the broader MIS planning process, and particularly why IT investment alternatives should be considered. It will also help to assure those stakeholders who must provide the financing and other support, that the firm has considered the alternatives to IT investment and should proceed to the next level of IT investment decision-making (i.e., note Figure 2, Step 5, "Analyze and evaluate project alternatives using IT investment methodology").

While IT outsourcing is very popular and can be used to avoid many of the IT investment decision-making pitfall situations this book seeks to address, the most common use of this strategy, as suggested in the selective IT scouring strategy above, is where the firm partially outsources and partially maintains its own IT resources. This being the case, the entire procedure for MIS planning, and in particular the IT project planning component will be needed at some time by all MIS managers.

Summary

The purpose of this chapter has been to undertake, through a variety of differing methodologies, the analysis required for the first four detailed steps in the MIS planning process for IT projects (note Figure 2). To accomplish this, we have redefined in greater detail the tactical steps in MIS planning that lead up to IT investment decision-making analysis. To complete these preliminary steps we have introduced needs analysis, reengineering, and outsourcing as prerequisite procedural methods.

Regardless of whether you choose any of the alternatives IT investment strategies mentioned in this chapter or if you indeed plan on making a major IT investment, you will have to perform the types of analyzes presented in this chapter as a preliminary step to determine the types and specifications of your IT needs.

To complete the 5th Step (note Figure 2) in the MIS planning process for IT projects requires a variety of different analyzes to evaluate alternative choices in terms of their potential contribution to the organization (i.e., their value) and decision choices when compared in terms of each other. Starting in the next chapter and for the rest of this book, we explore differing methods for evaluating and choosing alternative IT investments.

Review Terms

Business processes	Process flowchart
Business process reengineering (BPR)	Process map
Flowchart map	Reengineering management
IT outsourcing	Reengineering program
IT projects	Request for proposal (RFP)
IT project team	Ripple effect
Needs analysis	Selective IT sourcing
Needs report	Service center

Discussion Questions

1. Why is the "ripple effect" important in IT investment decision-making?
2. How it is possible to avoid the steps in this chapter's MIS planning process that deal with IT project decision-making?
3. Do we have to use "reengineering" to complete the MIS planning process?
4. Why do you think "IT outsourcing" is so popular?
5. If "IT outsourcing" is such a popular strategy, why would there be anything inhibiting its use?

Concept Questions

1. What are the two components of a "needs analysis"?
2. What do you expect to find in a "request for proposal" document?
3. What is a "reengineering program" and how is it related to IT decision-making?
4. What are the four alternative IT investment strategies? Explain each.
5. What are five of the benefits of "IT outsourcing"?
6. What are three of the problems associated with "IT outsourcing"?
7. What are four of the barriers that limit the use of "IT outsourcing"?
8. What are five of the guidelines recommended for the implementation of "IT outsourcing"?

References

Bhatt, G.D., "Exploring the Relationship Between Information Technology, Infrastructure and Business Process Re-engineering," *Business Process Management Journal*, Vol. 6, No. 2, 2000, pp. 139-163.

Bhatt G.D. and Stump, R.L., "An empirically derived model of the role of IS networks in business process improvement initiatives," *Omega*, Vol. 29, No. 1, 2001, pp. 29-48.

Chase, R.B., Jacobs, F.R. and Aquilano, N.J., *Operations Management for Competitive Advantage*, 10th ed., Boston, MA: McGraw-Hill, 2004.

Chiesa, V., Manzini, R., and Tecilla, F., "Selecting Sourcing Strategies for Technological Innovation: An Empirical Case Study," *International Journal of Operations and Production Management*, Vol. 20, No. 9, 2000, pp. 1017-1037.

Currey, W., *Management Strategy for IT*, London, UK: Pitman Publishing, 1995.

Earl, M.J., "The Risks of Outsourcing IT," *Sloan Management Review*, Vol. 37, No. 3, 1999, pp. 26-32.

Greaver, M.F., *Strategic Outsourcing*, New York: NY, AMA Publication, 1999.

Hill, C.W., and Jones, G.R., *Strategic Management: An Integrated Approach*. Boston, MA: Houghton Mifflin, 1992.

Kangas, K., *Business Strategies for Information Technology Management*, Hershey, PA: Idea Group Publishing, 2003.

Kendall, K.E., Kendall, J.E., Pfleeger, S.L., Connelly, L., Cortada, J.W., Fogler, H.S., LeBlanc, S.B., Woolever K.R., and Loeb, H.M., *Project Planning and Requirements Analysis for IT Systems Development,* 2nd ed., Boston, MA: Pearson, 2002.

Kendall, K.E., Kendall, J.E., Pfleeger, S.L., Connelly, L., Cortada, J.W., Fogler, H.S., LeBlanc, S.B., Woolever K.R., and Loeb, H.M., *Needs Assessment and Project Planning*, Boston, MA: Pearson, 2000.

Lacity, M.C., Willcocks, L. and Feeny, D., The Value of Selective IT Sourcing," *Sloan Management Review*, Vol. 37, No. 3, 1999, pp. 13-25.

Lackow, H.M., "IT Outsourcing Trends," *The Conference Board*, Research Report 1289-01-RR, 2001, pp. 1-15.

Laudon, K.C. and Laudon, J.P., *Management Information Systems: Managing the Digital Firm*, 8th ed., Upper Saddle River, NJ: Prentice Hall, 2004.

Michaud, C. and Theonig, J.C., *Making Strategy and Organization Compatible*, New York, NY: Palgrave/Macmillan, 2003.

Outsourcing Information Technology: Cutting Costs Or Cutting Your Throat? *National Underwriter /Life & Health Financial Services*, Vol. 107 Issue 27, 2003 (July 7), pp. 42-45.

Thibodeau, P., "Offshore's Rise Is Relentless," *Computerworld*, Vol. 37, No. 26, 2003, pp. 1-2.

Schniederjans, M.J. and Cao, Q., *E-commerce Operations Management*, Singapore: World Scientific Publishing, 2002.

Shriver, R.F. and Wold, G.H., *Information Systems Management and Technology*, Rolling Meadows, IL: Bankers Publishing, 1990.

Sislian, E. and Satir, A., "Strategic Sourcing: A Framework and a Case Study," *The Journal of Supply Chain Management*, Vol. 36, No. 3, 2000, pp. 4-11.

Sylla, C. and Wen, H.J., "A Conceptual Framework for Evaluation of IT Investments," *International Journal of Technology Management*, Vol. 24, Nos. 2/3, 2002, pp. 236-261.

Turban, E., Rainer, R.K. and Potter, R.E., *Introduction to Information Technology*, New York, NY: John Wiley & Sons, 2001.

Wheelen, T. and Hunger, D., *Strategic Management and Business Policy*, Upper Saddle River, NJ: Prentice-Hall, 2003.

Chapter 3

Measuring Information
Technology Performance

Learning Objectives

After completing this chapter, you should be able to:

- Define the "economics of information".
- Explain why we have to measure IT value of investments.
- Describe financial performance measures used in IT investment decision-making.
- Explain what measures of IT effectiveness are used for in IT investment decision-making.
- Explain what measures of IT efficiency are used for in IT investment decision-making.
- Explain some of the cost categories that should be considered in IT investments.

Introduction

Before one can evaluate an investment, the investment must be measured in some way for comparison. In this chapter we introduce a variety of theories and methods for the valuation of investments in IT. The main focus of this chapter is to present differing concepts on measuring the business value of IT from a variety of differing standpoints for consumers, and for internal and external IT sourcing agents. The chapter also focuses on the issues of measuring business effectiveness and efficiency.

The Economics of Information

The true beginnings of the study of *IT management* is unknown. It appears that managing and operating IT has been in existence since the 1960's (McNurlin and Sprague, 2004, p.2). It may be that IT management was important but simply did not appear in the literature until that time period. Clearly the cost savings afforded by these systems more than justified their use, irrespective of how they were managed.

A shift has taken place, as technology has evolved and improved, from a cost savings objective to one of generating revenue. Today, some of the specific objectives of IT are to improve customer satisfaction, reduce errors, obtain competitive advantage, and improve staff morale, all aimed at the overall objective of increasing revenues. The benefits afforded by IT under these types objectives tend to be very difficult to measure and manage, due to either their subtle effects or their intangible nature. IT benefit identification, measurement, and management may be considered some of the most important and difficult tasks for IT management.

Economics is the study of the allocation of scarce resources. Economics provides tools to analyze, evaluate, and select among competing alternatives. IT investment decision-making is in itself an economic decision. Scarce resources must be allocated in an optimal way, so as to select the best IT investment alternative. Economic tools such as supply, demand, costs, and benefits may be used in a variety of different ways in IT investment decision-making.

The *Economics of Information* refers to a systematic series of concepts and theories that explain the role which information and IT play to assist an organization with product and service design, development, manufacture, and delivery. The economics of information presents rules with which to assess the expected and actual impact of investing in a particular IT. Some work has been done to construct the economics of information for IT investments; however, few formal, widely accepted concepts, theories, and rules have come into existence to actually assist in the assessment of IT investments. One reason for this lack of theoretical development is that information and IT are different from the common types of goods and services already defined in economics. Digital books, digital libraries, computing networks, and digital publishing have different properties than common goods like steak and guns. Despite the lack of a formal economics of information, economics and its influence on IT investment decision-making is nonetheless important. Some of the more general economic theories of the past have been applied to the study of IT systems.

Examples of Economic Theories of Information

Borrowing from classic economics, the *transaction cost theory* is based on the logic that firms seek to economize on transaction costs. Originally proposed by Ronald H. Coase in 1937, this theory proposes that investments, in IT among other things, help reduce transactions costs and in turn reduce the size of the firm, making it more productive (Putterman and Kroszner, 1995). As we can see in Figure 1, a shift in transaction costs from "B" to "b", measured by transactions costs "A" to "C", results in a reduction in the size of the firm (i.e., employees, physical facilities, etc.) from "a" to "c".

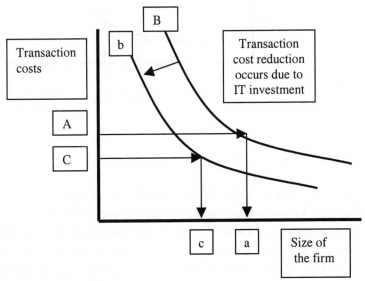

Figure 1. Transaction cost theory.

Another applicable economic theory is "agency theory" (Laudon and Laudon, 2004, p. 83). Agency theory can be viewed as dealing with the issue of the impact of IT on employees, or more aptly referred to as "agents" who work for the owners of businesses. This theory suggests that as firms grow in size, the owners have to increase the number of employees who work as agents for them to support the complexity of the organization as its networks and general size dictate increased activity. By investing in IT that saves time and improves the control of the firm for the owners, less agents or employees are required to manage the firm. As shown in Figure 2, these agency cost reductions (from "A" to "C") move the agency costs curve from "B" to "b", causing a reduction in the size of the firm, principally a reduction in employees.

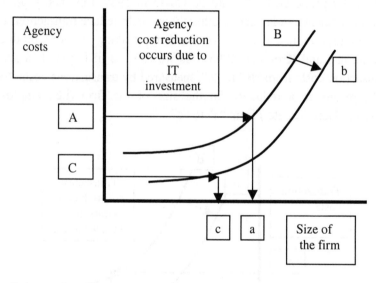

Figure 2. Agency theory.

An IT budget must be allocated to competing alternatives investments in computers, digital storage space, bandwidth, and even full-blown systems. Each IT investment alternative affords different benefits to various stakeholders, and economic theories, concepts, and practices may be used to help quantify benefits and to determine the best allocation of an IT budget.

Why Measure IT Performance?

Identifying, measuring, and managing IT benefits and costs are some of the most important and difficult of tasks for IT managers. These tasks must be undertaken so that an overall performance measurement system may be put into place and used to evaluate the functioning of an IT investment. *Performance measurement* allows decision-makers to assess the business value, and the efficiency and effectiveness of an IT. A *performance measurement system* evaluates the effects of an IT and may be used to justify an initial IT investment and later to assess its impact after implementation and use. Measuring IT performance provides decision-makers with facts that may be used to allocate resources in the optimal way. It may be valuable to conduct performance measurement at several different times throughout the lifecycle of an IT and to do so at different levels of the organization.

Measurement is vital to fully understand the impact of IT investment on overall organizational performance. In the continuous cycle of planning and control, measurement allows organizations to monitor and manage IT performance. Tracking measures like number of time periods to deliver an IT, mean-time-to-failure, number of training days per time period, and revenue growth to name a few, allows management to make assessments, adjustments, and decisions with respect to IT investment and ultimately IT strategy. An effective measurement program allows the organization to monitor costs, make good decisions with respect to the allocation of IT resources and to develop improvement strategies. More and more organizations are relying on IT to facilitate the supply of their goods and services and as this reliance increases, so will the need for an IT measurement program. IT, like any other aspect of business, needs to be monitored and managed, and a measurement program is an effective way of doing so.

Measurement of IT performance is essential for a variety of other reasons. Organizations invest major portions of their budgets on IT. It is not unusual for some firms to allocate 50 percent of all corporate investment to IT investment. Management is often concerned about the return from these large investments and measurement is one way to justify that IT contributes positively to overall organizational performance. Measurement also provides useful information to evaluate and prioritize different IT projects. Deciding which IT investment to actually make can be very difficult and measurement assists by providing information about the potential return from an investment and thus a way to prioritize investments. Another reason to measure IT performance is that it improves communication between executives and IT suppliers, whether it is an internal IT department or an outside supplier. The communication between IT professionals and management from other parts of the organization has traditionally been problematical. It seems that IT professionals tend to "speak a different language" than other members of management. Measuring IT performance allows IT professional to speak in the same terms, such as profit, loss, and return on investment, as other members of management, hence improving communication. Yet another reason to measure IT performance is that the whole measurement and evaluation process is a valuable learning experience. The knowledge gained from designing, implementing and utilizing an effective measurement program can be invaluable for truly understanding the effects of IT investments.

IT investments may be considered unique types of investments because they typically have both tangible and intangible effects on an organization. Because of the tangible and intangible nature of potential and realized effects of IT investments, both objective and subjective measures are necessary. For example, suppose the major benefit of an IT investment is to improve the quality of information for management decision-making. It may be necessary and more appropriate to use a subjective measure, like management opinion surveys, to measure the effect of this IT investment benefit on organizational performance. In addition, objective measures, like increased revenue, may also be used to assess the impact of this IT investment on organizational performance. In most cases, both objective and subjective measures are necessary to account for the tangible and intangible nature of IT effects. As a result, an effective measurement program will include both objective measures (e.g., costs, profit, revenue, etc.) and subjective measures (e.g., customer ratings, performance rankings, manager opinion scales, and user satisfaction surveys, etc.) to assess the impact of IT on organizational performance.

An effective measurement program will also measure performance at appropriate levels of an organization. IT is often so intertwined with all aspect of a business that it affects at different levels, functions and business processes within an organization. Due to this type of impact, measurement of IT effects must be conducted at all appropriate organizational levels. It is recommended that, in general, measurement occurs at the business level by assessing the business value of IT and the overall effectiveness of IT. At lower levels in the organization measurement can be used to assess the effectiveness and efficiency of IT and of the IT sourcing function. The remainder of the chapter is focused on exploring the significance of these aspects of IT measurement.

Measures of IT Business Value

The business value of IT maybe defined as the overall value of IT for a particular organization. Assessing the business value of IT attempts to provide insights into the effects IT investments have on the bottom line performance of an organization. IT investments may contribute to overall organizational performance by improving the financial performance, business performance, and strategic performance of an organization.

The first step to determine the business value of IT is to identify objectives of IT investments in each of the previously mentioned three areas. Once objectives have been identified, one must select at least one measure, preferably more than one, to assess each of the objectives. As will be seen in the following sections, several measures exist for each of the three areas, however, only the most appropriate measures should be selected and used to evaluate the business value of IT.

Financial Performance Measures of IT

To evaluate the business value of IT, the financial performance of the organization should be assessed. Typically, measures to assess the financial performance of an organization include return on investment ratios like *return on equity* (ROE), *return on assets* (ROA), *earnings per share* (EPS), *return on sales* (ROS), and *return on investment* (ROI). Table 1 shows these and other common measures that may be used to evaluate the financial performance of an organization. Table 1 also provides the basic calculation and interpretation of each. It should be noted that this list is not exhaustive and that other measures exist and may be used; however, the ones in Table 1 are commonly used to determine financial performance of an organization. Note also that there might be more than one acceptable way to calculate the ratios.

In addition to the common financial ratios, more specific IT investment ratios may be used to assess the business value of IT. Some of these measures are presented in Table 2. These are common measures used by firms that are members of *Computerworld's Premier 100* (Sethi, Hwang and Pegels, 1993). Computerworld developed an index to rate the overall effectiveness of IT organizations. The index uses five IT investment measures: (1) IT budget as a percentage of revenue; (2) Market value of an organization's IT equipment as a percentage of revenue; (3) Percentage of IT budget spend on IT staff; (4) Percentage of IT budget spend on IT staff training; (5) Number of personal computers and terminals as a percentage of total employees. The IT investment ratios, as well as the financial performance ratios may be tracked overtime and be compared to benchmark ratios of other competitors and industry averages. This type of analysis assists organizations in determining the financial performance of IT.

Assessing the financial performance of IT may not necessarily result in an overall number that can be used to measure the performance of IT. The knowledge gained from assessing the financial performance of IT can be used to gauge the business value of an IT investment or the overall business value of IT.

Table 1. Financial performance measures.

Ratios	Calculation	Explanation
Profitability measures		
Return on equity	Income available to common shareholders from continuing operations divided by common shareholder's equity	Measures profitability of the investment to the owners
Return on assets	Income available to common shareholders from continuing operations divided by average total assets	Measures profitability and how efficiently assets were utilized
Return on investment	Income available to common shareholders from continuing operations divided by total invested capital	Measure profitability based on total investment, both debt and equity
Return on sales	Income available to common shareholders from continuing operations divided by net sales	Measures profitability based on sales
Earnings per share	Total earning divided by total shares outstanding	Measures earnings per share value
Revenue growth	Revenue for the current period minus revenue from the prior period divided by revenue from the prior period	Measures growth in revenue over the prior period
Efficiency measures		
Sales by total assets	Net sales divided by average total assets	Measures how efficiently assets were used to generate sales
Sales by employee	Net sales divided by number of employees	Measures sales ability effectiveness
Inventory turnover	Cost of goods sold divided by average inventory	Measures the liquidity of inventory and how fast it is sold

If an IT is supporting organizational objectives and those objectives are intended to increase revenue then it may be inferred that the IT is valuable. This value is dependent on the importance of objectives an IT supports. If an IT supports extremely important objectives that have a great effect on the bottom line, then its value is greater than another IT that supports a less important or influential objectives. For example, to actually measure the benefit of improved customer satisfaction and the exact dollar amount of increased sales due to improved customer satisfaction may be impossible. However, through a survey it may be determined that customers report they are more satisfied because of an IT and buy more product because they are more satisfied. As a result, it may be inferred that the IT improved marketing performance.

Business Performance Measures of IT

"Business performance" is affected by many internal and external factors, including the implementation and use of an IT. *Business performance* is the result of the execution of business processes and the allocation of resources to business processes. Business performance depends on how well a business allocates its resources to business processes and how well it performs its business processes. In these situations the role of IT is to support business processes. As a result, IT may be viewed as contributing to the business performance of an organization indirectly by supporting the business processes that contribute to its business performance.

It may be most appropriate to assess business performance with both financial and non-financial indicators. The "balanced scorecard" may be a very practical way to measure the business performance of IT and incorporate both financial and non-financial measures. The *balanced scorecard method* is a performance measurement tool that allows organizations to assess the performance of the business as whole, individual departments within the organization and specific projects. It was developed by Kaplan and Norton and was initially presented in the *Harvard Business Review* (Kaplan and Norton, 1992). The balanced scorecard provides a balanced representation of financial and non-financial performance through four perspectives: (1) the customer perspective (note Table 2); (2) the internal business perspective (note Table 3); (3) the innovation and learning perspective (note Table 4); (4) the financial perspective (note Table 5). Objectives (goals) are identified for each perspective and then performance measures are selected to measure each objective. We will describe the balanced scorecard method more fully in Chapter 7.

Table 2. Business performance measures from the customer perspective.

Objective	Examples of measures
Cost effectiveness	Processing cost per some unit of measure
Cost effectiveness	Data center costs per workstation
Delivery speed	Number of time periods to deliver product
Support availability	Mean-time-between-failures
On-time delivery	Number of times actually met delivery schedule
Defective products/services	Number of defects
Training support	Percent of user groups training is held regularly
Responsive support	Number of time periods to fix problem

Table 3. Business performance measures from the internal perspective.

Objective	Examples of measures
Efficient development	Number of staff days per project by size
Quality development	Number of defects in test unit
Low levels of rework	Number of times software was reworked
Competent employees	Number of years by job class
Highly skilled employees	Number of hours of training per employee

Table 4. Business performance measures from the innovation and learning perspective.

Objective	Examples of measures
Highly skilled employees	Number of training days per time period
Protect knowledge based on employees	Number of employee resignations
Increased market offerings	Number of new product introductions per time period
Encourage innovative research	Percent of budget spent on IT research
Improve innovative thinking	Number of employee ideas implemented
Foster innovative research	Number of state of the art projects per time periods

Table 5. Business performance measures from the financial perspective.

Objectives	Examples of measures
Profitable business	Revenue growth
Profitable business	Return on sales or assets
Profitable use of IT	Revenue per user group
Profitable IT development	Profit per employee
Efficient maintenance	Cost of IT maintenance as a percent of some unit of measure
Reliable planners	Adherence to budget

Strategic Performance Measures of IT

The third area in which IT investments may contribute to overall organizational performance is through the improvement of the strategic performance of an organization. Strategic performance depends largely on a few key areas that the organization must excel at to survive and thrive. These areas, called *critical success factors* (CSF), are the limited number of areas in which satisfactory results will ensure successful competitive performance for the organization. CSFs are the few key areas where things must be done well for the organization to flourish. We will discuss CSFs in Chapter 7. The CSF method has been used for IT investment decisions as reported by Slevin, *et al.* (1991), Boynton and Zmud (1984) and Munro (1983).

Costs of IT

To assess the business value of IT, information must be collected as to the actual amount spent on IT. IT costs maybe difficult to track because more individual departments are being charged with the task of planning and managing their own IT needs. IT costs in a central IT budget will be relatively easy to collect, compared to those within departmental budgets. However, all IT costs must be collected to determine the true costs of IT to be used in determining the business value of IT. Once the actual costs of IT have been identified, then they may be categorized in some way and this data compared to that of competitors. Some common categorizations of IT costs are presented in Table 6.

Table 6. Cost categorizations for IT.

Costs by Activity	Costs by Resource
1. Development costs	1. Technology costs
2. Maintenance costs	2. Personnel costs
3. Operating costs	3. Outside services costs
4. User support costs	4. Other costs
5. Administration and other costs	

The IT cost data may be used in the analysis of the business value of IT. Van der Zee (2002) suggests using the following measures to support and complement the measures to assess the business value of IT: IT costs as a percentage of revenue, growth rate of IT budget, IT spending by resource, and IT spending by activity. The measures are easy to calculate and compare to industry data, however, it is cautioned by Van der Zee (2002) that this cost analysis has not been proven and should never be used in isolation to make decisions as to the business value of IT.

Measures of IT Effectiveness Value

IT effectiveness value may be defined as the extent to which IT supports business processes and employees. Effectiveness is associated with "doing the right things", meaning that management delivers the right products and services, with the right features, to the right customers, at the right time. Measuring the effectiveness of IT translates into measuring if the organization is making the right decisions with respect to IT investments. In other words, IT managers should question whether or not their organization is making the right IT investments to support business processes that deliver the right products and services, with the right features, to the right customers, at the right time. It seems that IT is able to support many of the business processes, activities, and tasks performed by an organization. Some examples of such IT are customer relationship management systems, inventory management systems, production scheduling systems, decision support systems and executive support systems, to name a few. Since ITs are intertwined with every aspect of the business, measuring the effectiveness of such systems is essential to determine their impact on overall organizational performance.

To determine the effectiveness of IT one may evaluate the following three aspects of a business: (1) the extent to which IT supports business processes; (2) the extent to which IT supports employees; and (3) the extent to which the IT sourcing function meets business requirements. To evaluate the extent to which IT supports business processes, one should first identify the business processes performed by the organization and the different types of IT. A team of employees and management may identify, describe and list the different business processes the organization uses to create and deliver their products/services. Porter's (1985) *value chain model* may used to facilitate this task as a framework with which to identify specific business processes in each of the more general categories of inbound logistics, operations, outbound logistics, marketing, sales, service, firm infrastructure, human resource management, technology development, and procurement. After identifying the business processes, a team of

Table 7. Categorizations of IT.

By IT purpose (Ross and Beath, 2002)	By IT purpose (Remenyi, Money, and Sherwood-Smith, 2000)	By management objective (Weill, 1992)	By IT purpose (Davenport, 1993)
Process improvement	Must do	Strategic	Automational
Renewal	Vital/core	Informational	Informational
Experiments	Critical/core	Transactional	Sequential
Transformation	Strategic/prestige	Threshold IT	Tracking
	Architectural/must do		Analytical
			Geographical
			Integrative
			Intellectual
			Disintermediating

employees and management should identify the different categories of IT. The IT categorization may be unique to each individual organization; however, different categorizations have been put forth in the literature and several are presented in Table 7.

After categorizing IT, one should create a *process/IT matrix* of business processes mapped against the types of IT and identify the current IT application(s) in each cell of the matrix as presented in Figure 3. The IT applications should include hardware, software, and systems whose IT purpose is utilized in some way, to meet the informational or processing needs of the various processes.

This allows the team of employees and managers to determine which areas of the business are supported by IT and which are not. Analysis of the matrix will also assist with determining which business processes should be supported by IT (i.e., technological advances may create opportunities to support business processes that could not have been supported in the past). The last step is to create and calculate an effectiveness measure. This business process matrix technique allows the assessment of the "coverage" of IT. A coverage ratio may be calculated as the actual applications divided by the potential applications for each cell in the matrix. This ratio may be used to determine which business processes need "more coverage." Table 8 presents the IT effectiveness measures (the coverage ratios) for IT support of business processes.

Processes	**Types of IT**	Process improvement	Renewal	Experiments	Transformation
Processes					
Firm infrastructure					
Formulate data policies					
:					
Design work processes					
Procurement					
Conduct audits					
:					
Provide security					
Operations					
Provide quality assessment					
:					
Respond to Help Desk needs					

Figure 3. Example of process/IT matrix.

Table 8. Effectiveness measures for IT to support business processes.

IT effectiveness factor	IT effectiveness criteria	IT effectiveness measure
Coverage	Coverage of management business processes	Actual application of IT as a percent of potential per IT type
Coverage		

Coverage of operational business processes	Actual application of IT as a percent of potential per IT type

In summary, the following steps should be undertaken to assess one aspect of the effectiveness IT, more specifically, the extent to which IT supports business processes:

1. Identify business processes;
2. Identify types of IT;
3. Create a matrix of business processes and types of IT and identify the current IT application(s) for each cell in the matrix and
4. Calculate the ratio of actual applications to potential applications for each cell.

Qing and Plant (2001), Sylla and Wen (2002) and Van der Zee (2002) have identified a number of general effectiveness factors for IT, some of which are presented in Table 9. These effectiveness factors represent the general categories of actual effectiveness factors that may be used to identify effectiveness criteria and effectiveness measures. The effectiveness factors are similar to the objectives of an IT. Notice in Table 9, coverage is an IT effectiveness factor. From the previous discussion, the IT coverage of business processes is just one way to assess the effectiveness of IT. Organizations may select differing appropriate IT effectiveness factors and their corresponding IT effectiveness criteria, then use them to devise effectiveness measures.

Table 9. IT effectiveness factors.

Accuracy	Operability
Availability	Portability
Connectivity	Reliability
Comprehensibility	Reparability
Coverage	Responsiveness
Flexibility	Reusability
Instability	Robustness
Integrity	Security

Learnability	Testability
Maintainability	Understandability
Manageability	User-friendliness

Table 10. Effectiveness measures for IT to support employees.

IT effectiveness factor	IT effectiveness criteria	IT effectiveness measure
Reliability	Reliability of IT applications	Mean times between failures and to repair; failures per 100 hours of operation
Reliability	Reliability of information	Correct data as a percent of total data available
Accessibility	Accessibility of information	Mean response time, batch turnaround time
Security	Security of information	Number of secured data sets as a percent of total
Flexibility	Flexibility of IT	Time required to make requested changes
User-friendliness	Ease of use	User-friendliness rate on a ratio scale

To evaluate the second aspect of IT effectiveness, the extent to which IT supports employees, a similar conceptual model to that used to identify measures for the business process aspect may be employed. The first step here is to identify "IT capabilities". *IT capabilities* are everything and anything an IT can accomplish. Again, a team of employees and managers will identify IT capabilities. The second step is to define effectiveness criteria and effectiveness measures for each IT capability. Examples of effectiveness criteria and effectiveness measures are presented in Table 10.

In summary, the following steps should be undertaken to assess the second aspect of the effectiveness IT, more specifically, the extent to which IT supports employees:

1. Identify IT capabilities;
2. Define effectiveness criteria for all IT capabilities; and
3. Develop measures for effectiveness criteria.

The third aspect of assessing IT effectiveness is to measure the extent to which IT sourcing meets business requirements. The previous discussion of IT effectiveness has focused on measuring the effectiveness of the actual IT. It is also appropriate to measure the effectiveness of the IT sourcing function, whether internal or external to the organization. The *effectiveness of IT sourcing* may be defined as the extent to which the IT sourcing function delivers IT products and services that support business requirements. This evaluation should focus on how well the IT department or an outside IT provider is supplying IT that is used to meet business requirements. The same conceptual model for evaluating the extent to which IT supports employees may be used to develop effectiveness measures for the extent to which IT sourcing meets business objectives, with one modification: use IT sourcing capabilities instead of the actual IT capabilities. Table 11 shows some examples of IT effectiveness criteria and corresponding effectiveness measures.

Table 11. Effectiveness measures for IT sourcing.

IT effectiveness factor	IT effectiveness criteria	IT effectiveness measure
Operability	Ease of operation of IT	Number of outages, file recoveries, incidents; ease of operation, rated on a ratio scale; mean time to repair
Maintainability	Ease to repair	Mean time/effort to repair/adapt/test: quality of documentation rated on a ratio scale
Flexibility	Ease with which maintenance can be performed	Time to perform maintenance
Testability	Ease with which to test IT	Time it takes to test IT
Reusability	Extent to which IT parts can be re-used	Number of components reused
Portability	Ease to transfer IT capability to another application	Mean time/effort to transfer IT components
Connectivity	Ease to link one IT capability with another	Number of IT components not adhering to standards as a percent of total

Table 11 (Continued)

Security	Extent to which IT meets security necessities	Secured data set as a percent of total
Scalability	Ease of expansion	Amount of time required to make improvements

An alternative method for identifying and selecting IT sourcing effectiveness measures is the balanced scorecard method. A set of four scorecards may be developed for each of the identified functions of an IT department or provider. The set of four scorecards will each individually represent one of the four perspectives from the customer, operational, innovation and learning, and financial perspectives. The first step is to identify the basic functions of an IT department or provider. Typical functions include IT infrastructure management, IT development management, and client support. If the IT department is a profit center, then the function of sales and marketing may be included as well as an overall function of IT supply management. The next step is to identify the objectives for each function, from each perspective. A list of possible effectiveness measures is presented in Table 12.

It should be noted that a large number of measures exist to assess the effectiveness of an IT and that the organization must select appropriate measures from the set of proposed ones with respect to the particular organization, situation and IT investment. The measures presented in this chapter are examples of the many that may be used to assess the business value, effectiveness, and efficiency of IT.

Measures of IT efficiency value

Efficiency refers to producing the desired effect with minimum effort and resources. *IT efficiency* can refer to producing the desired effect with minimum IT expenses. The efficiency of IT is usually not assessed, because IT is only one part of the many cost components that affect the efficiency of an organization. In addition, to determine the efficiency of an IT, one would be required to allocate IT expenses from IT infrastructure, mandatory IT, and IT research investments. This allocation may be nearly impossible because these types of investments support multiple business processes of the organization. As such, the efficiency of the IT sourcing function or provider is the only facet of efficiency that will be assessed in this chapter.

The *Efficiency of IT sourcing* may be defined as the extent to which effective IT is supplied at minimum cost. By this definition, IT sourcing must first be effective. Only after IT sourcing is "doing the right things"

Table 12. Additional effectiveness measures for IT sourcing.

Objective	IT Effectiveness Measure
Be an attractive supplier	Overall client/user satisfaction score
Be a good employer	Employee satisfaction score
Be a reliable planner	Performance to budget; percent of project delivered on time, within budget
Be a responsive supplier	Number of hours/days to fix problem or make change
Be quality developer	Number of defects in unit test and system integration by test size
Be competent	Number of years experience by job class
Be a quality operator	Number of outages/defects/incidents
Cultivate innovation	Number of training days per year; percent of total time devoted to training
Be available	Percent of inquiries answered per day
Be informed	Number of informational meetings attended per year
Create new markets	Percent of revenues from new applications, products and/or relationships

can efficiency be fully gained. It may be very costly to become effective so complete efficiency may not be experienced by most organizations. In addition, the definition of efficiency allows for different interpretations of what is actually efficient. An efficiency level in one organization may be considered "very efficient", and in another organization "moderately efficient" assuming the IT function is equally effective. Both efficiency and effectiveness are relative measures that must be compared to a benchmark level.

Just as for the effectiveness of the IT sourcing, the balanced scorecard method may be utilized to obtain measures to assess the efficiency of the IT sourcing function. Again, a list of functions of the IT sourcing department or provider must be decided upon, with common ones being IT infrastructure management, IT development management, client support, sales and marketing, and the overall function of IT supply. The next step is to identify the objectives for each function, from each perspective. The balanced scorecard provides a comprehensive view of the IT department or supplier by analyzing the customer, internal, innovation and learning, and financial perspectives of the department. Once objectives have been identified, appropriate measures can be selected or developed to measure the extent to which the IT department or provider is meeting the objectives. Some potential measures of the efficiency of IT sourcing are presented in Table 13. For an in-depth discussion of applying the balanced scorecard approach to evaluate the IT sourcing department or provider see Van der Zee (2002).

Summary

This chapter has overviewed concepts of economics and measurement metrics useful in the evaluation of IT performance, efficiency, and effectiveness. The purpose of this chapter was not to provide tacit decision models, but acquaint you with concepts and measurement issues on the criteria and input into models that will be presented in later chapters.

This chapter completes Part I and provides a basic foundation for the remaining chapters in this textbook that introduce specific valuation methods that use the various criteria as inputs into more complex investment decision models. In Part II "Financial Information Technology Investment Methods," consisting of the next three chapters of this textbook, we begin our examination of the investment methodologies with some of the more traditional financial investment methods.

Table 13. Efficiency measures for IT sourcing.

Objective	IT efficiency measure
Handle costs successfully	Total development costs per delivered Function Point
Handle costs successfully	Processing costs per on-line business transaction
Handle costs successfully	Data center costs per business user/workstation
Handle costs successfully	Costs of marketing programs, events, promotion material, etc., as a percent of revenue
Optimize asset utilization	CPU usage overall, prime shift, non-prime shift
Optimize staff levels	System programmers per Operating System/used MIPS
Be a quick adopter	Average elapsed time to master new development approaches/techniques/tools
Be an efficient developer	Number of staff days by project size, by project phase
Be an efficient supplier	Number of users/workstations supported per staff member
Be in operational control	Performance to budget by application/user group

Review Terms

Agency theory
Balanced scorecard method
Business performance
Business value of IT
Critical success factors (CSF)
Earnings per share (EPS)
Economics
Economics of Information
Effectiveness of IT sourcing
Efficiency
Efficiency of IT Sourcing
IT capabilities

IT effectiveness value
IT efficiency IT management
Performance measurement
Performance measurement system
Process/IT matrix
Return on assets (ROA)
Return on equity (ROE)
Return on sales (ROS)
Return on investment (ROI)
Transaction cost theory
Value chain model

Discussion Questions

1. How are the "economics of information" useful in IT investment decision-making?
2. Why measure IT performance?
3. Why is "business performance" so important in IT investment decision-making?
4. What is the difference between the profitability and efficiency measures in the context of "financial performance measures"?
5. Why is "IT effectiveness" important in IT investment decision-making?
6. Why is it important to have IT effectiveness measures that support employees?

Concept Questions

1. What is "transactions cost theory"? How it is related to IT?
2. What is "agency theory"? How it is related to IT?
3. What is "business performance"? How is it measured?
4. What are four "financial performance measures"? Explain what each seeks to do.
5. Why is it necessary to collect cost information in order to assess the business value of IT information?

6. What are "IT effectiveness values"?
7. How do we use the "value chain model" in IT investment decision-making?
8. What are "IT effectiveness factors" and how are they used in IT investment decision-making?
9. What are the "effectiveness measures for IT sourcing"? Listing any two objectives, what "IT effectiveness measures" can be used for IT sourcing?
10. What are the "efficiency measures for IT sourcing"?

References

Boynton, A.C., and Zmud, R.W., "An Assessment of Critical Success Factors," *Sloan Management Review*, Vol. 25, (Summer 1984), pp. 17-27.

Davenport, T.H., *Process Innovation: Reengineering Work through Information Technology*, Boston, MA: Harvard Business School Press, 1993.

Laudon, K.C. and Laudon, J.P., *Management Information Systems: Managing the Digital Firm*, 8th ed., Upper Saddle River, NJ: Prentice Hall, 2004.

Lee, C.S., "Modeling the Business Value of Information Technology," *Information & Management*, Vol. 29, 2001, pp. 191-210.

McNurlin, B.C. and Sprague, R.H., *Information Systems Management in Practice*, 6th ed., Upper Saddle River, NJ: Person/Prentice Hall, 2004.

Munro, M.C., "An Opinion ...Comment on Critical Success Factors Work," *MIS Quarterly*, Vol. 7, (September 1983), pp. 67-68.

Porter, M., *Competitive Advantage*, New York, NY: Free Press, 1985.

Putterman, L. and Kroszner, R., *The Economic Nature of the Firm: A Reader*, Cambridge, UK: Cambridge University Press, 1995.

Remenyi, D., Money, A. and Sherwood-Smith, M., *The Effective Measurement and Management of IT Costs and Benefits*, 2nd ed., Oxford, MA: Butterworth-Heinemann, 2000.

Ross, J.W. and Beath, C.M., "Beyond the Business Case: New Approaches to IT Investment," *MIT Sloan Management Review*, (Winter 2002), pp. 51-59.

Sethi, V., Hwang, K.T. and Pegels, C., "Information Technology and Organizational Performance, A Critical Review of Computerworld's Index of Information Systems Effectiveness," *Information & Management*, Vol. 25, 1993, pp. 193-205.

Slevin, D., Stieman, P.A. and Boone, L.W., "Critical Success Factor Analysis for Information Systems Performance Measurement and Enhancement," *Information & Management*, Vol. 21, 1991, pp. 161-174.

Sylla, C. and Wen, H.J., "A Conceptual Framework for Evaluation of IT Investments," *International Journal of Technology Management*, Vol. 24, Nos. 2/3, 2002, pp. 236-261.

Qing, H. and Plant, R., "An Empirical Study of the Casual Relationship Between IT Investment and Firm Performance," *Information Resources Management Journal*, Vol. 14, No. 3, 2001, pp. 15-26.

Van der Zee, Han, *Measuring the Value of Information Technology*, Hershey, PA: Idea Group Publishing, 2002.

Weill, P., "The Relationship Between Investment in Information Technology and Firm Performance: A Study of the Valve Manufacturing Sector," *Information Systems Research*, Vol. 3, No. 4, 1992, pp. 307-333.

Part II

Financial Information Technology Investment Methods

Chapter 4

Basic Financial Methods

Learning Objectives

After completing this chapter, you should be able to:

- Describe "basic financial methods" of IT investment.
- Use "breakeven analysis" for IT investment decision-making.
- Use "payback period method" for IT investment decision-making.
- Use "accounting rate of return" for IT investment decision-making.

Introduction

Everyone who makes an investment in information technology (IT) will be expected to perform a convincing analysis to justify their proposed investment. In the previous chapters we have focused on many issues that will help begin the preliminary steps in that analysis. This chapter is the first of several chapters focusing on fundamental financial components or the "basic financial methods" expected in IT investment decision-making analysis.

What are the Basic Financial Methods?

Financial methodologies, in general, are rooted in subject areas of Finance and Accounting, and have been used in traditional "capital

81

budgeting decisions" for many years. *Capital budgeting decisions* are those concerning investment in real assets, e.g., machinery, facilities, management expertise and information technology. *Real assets* tend to be long-lived assets, meaning they will be used for a long period of time, often several years. However, the useful life of an IT investment tends to be shorter than that of the typical real asset (e.g., a facility or piece of machinery is often used for several decades where as information system may become obsolete in two to three years). Capital budgeting techniques are commonly used in IT investment decision-making and it is assumed that an IT that produces benefits beyond one year is considered to be a long-lived asset. In other words, capital budgeting techniques may be employed because an IT investment produces benefits beyond one year and its useful life extends into the long-term future.

Financial methodologies may be divided into two categories: (1) basic financial methods; and (2) advanced financial methods. *Basic financial methods* are those most often employed because of their simplicity and include breakeven analysis, payback period methodology and accounting rate of return methodology. Basic financial methods are easy to calculate, understand and communicate to others. Their simplicity appears to be their attraction. There are other methods, considered *advanced financial methods.* Some of these are presented in Chapter 5 of this textbook and include present value analysis, profitability index, return on investment and internal rate of return methodologies. The calculation for these techniques is more involved and, for several of these methodologies, requires knowledge and understanding of the time value of money and present value. Present value analysis involves discounting cash flows received in the future to their present value so the computation is more difficult than that of the more basic financial methods.

In traditional capital budgeting decisions, evaluating common *capital assets* such as equipment and facilities, the "payback period" methodology has been the most widely used financial methodology (Sangster, 1993). It appears that this pattern of use has carried over to IT investment decision-making. It has been shown that the payback period methodology is again the most widely used financial methodology to evaluate IT investment projects (Ballantine and Stray, 1999, Ballantine and Stray, 1998, Bacon, 1992 and Tam 1992). Results of these studies, as shown in Table 1, indicate that the payback period is the most extensively used financial methodology, while the discounted cash flow methods are less frequently utilized in IT investment evaluation.

Table 1. Use of financial methodologies in IT evaluation.

	Ballantine and Stray 1999 (%)	Ballantine and Stray 1998 (%)	Bacon 1992 (%)
Payback	69	60	61
ARR	48	43	16
NPV	15	24	49
IRR	15	25	54

Ballantine and Stray (1999; 1998) show that 69 and 60 percent, respectively, of the organizations surveyed used payback period methodology on their most recent IT investment. Bacon (1992) found that 61 percent used of the companies surveyed used payback period as one criterion to evaluate IT investments. Tam (1992) found that simple financial techniques, like payback period, are preferred by organizations than complex discounted cash flow methods. It seems that the simplicity and familiarity of the payback period method may be major reasons for its extensive use in IT investment decision-making.

What is Breakeven Analysis?

Breakeven analysis for IT investment involves the comparison of quantifiable costs with quantifiable and non-quantifiable benefits of an IT. The breakeven point is where the total value of benefits equals that of total costs. IT investment alternatives that breakeven or that have benefits greater than costs are considered to be good investments and should be undertaken. *Non-quantifiable benefits* are referred to as intangible, meaning that a monetary value cannot be assigned to them. Breakeven analysis utilizes subjective assessment of intangible benefits and, as a result, is only used if for some reason the other more objective methods cannot be used. Breakeven analysis can be used to evaluate a single IT investment and to select one or several among a set of IT investments.

The first step in breakeven analysis is to calculate the present value of costs, assuming that the costs and benefits have been identified and valued. *Present value* (PV) is the current value of a cash flow that is to be received sometime in the future. PV analysis is based on the assumption that a dollar today is worth more than a dollar tomorrow because it can be invested and begin accruing interest today. A more in-depth discussion of PV may be found in Chapter 5.

The next step is to calculate the present value of all quantifiable benefits and to subtract this value from the PV of costs. The result is referred to as *net costs*. The value of intangible benefits must equal net costs for the IT investment to breakeven. If the intangible benefits equal or exceed net costs then according to this methodology, the investment should be profitable and, thus, should be undertaken.

As calculated, net costs are the total net costs for the entire life of the investment. It may be beneficial to adjust net costs to reflect the yearly value of the intangible benefit, which is done by calculating the equivalent annual value. The final step in breakeven analysis is for management to subjectively determine if the intangible benefits are in fact worth at least the value of net costs. If they are, the investment should be undertaken, and if not the investment should not be undertaken.

Let's illustrate this with an example. Suppose that an organization must decide whether or not to invest in a computer system. The costs and benefits of the proposed system have been evaluated and are presented in Table 2. The *initial cost* of the computer system is $100,000 with an *annual cost* of $20,000 for three years, which includes such things as maintenance, service, training, and personnel. The anticipated benefits are improved productivity with a value of $35,000 per year for three years and improved employee morale, which is intangible.

Table 2. Costs and benefits of a computer system example.

	Year 0	Year 1	Year 2	Year 3
Costs:				
Initial cost ($)	100,000			
Annual cost ($)		20,000	20,000	20,000
Benefits:				
Improved productivity ($)		35,000	35,000	35,000
Improved employee morale		--	--	--

The PV of costs and tangible benefits are presented in Table 3. The net costs associated with this IT investment are $63,973, calculated as the PV of costs less the present value of tangible benefits ($148,037-$84,064). The value of intangible benefits, in this case improved employee morale, must be equal to or greater than the value of net costs for the investment to be undertaken. The value of net costs may be converted into a yearly value by calculating the equivalent annual value. The equivalent annual value is the constant amount per year whose present value over the three years is equal to the net costs of $63,973. That value is $ $26,635. Management must subjectively decide if the benefit of improved employee morale is actually worth $26,635 per year for three years.

Table 3. Present value calculations for the computer system example.

Type of value	Calculations
Present value of costs ($)	$PV = 100,000 + \dfrac{20,000}{(1+.12)^1} + \dfrac{20,000}{(1+.12)^2} + \dfrac{20,000}{(1+.12)^3} = 148,037$
Present value of tangible benefits ($)	$PV = \dfrac{35,000}{(1+.12)^1} + \dfrac{35,000}{(1+.12)^2} + \dfrac{35,000}{(1+.12)^3} = 84,064$

One way to determine the value of an intangible benefit is to survey well-informed managers. Managers who are familiar with the proposed computer system and the benefits and costs of the system may be able to assign a fairly accurate value to an intangible benefit. In this example, managers may be polled to determine if the value of improved employee morale due to the computer system is worth at least $26,635 per year for the next three years. If it is, then the organization should invest in the computer system and if it is not, then it should not invest.

Another way to determine the value of an intangible benefit is to use a surrogate measure or a group of surrogate measures that reflect the actual value. Managers may determine that improved employee morale may be measured by employee turnover, absenteeism, and/or higher quality work. These surrogates allow the intangible benefit of improved employee morale to be quantified and compared to the value of net costs. When suitable surrogates can be employed, these may be used in the initial present value calculations to determine if the IT investment breaks even.

The main advantage of breakeven analysis is that it allows intangible benefits to be considered in the analysis. The main disadvantage is that intangible benefits are subjectively valued and this value may not reflect their true value. The error in subjectively valuing a benefit may be large and there is no way to determine the amount of error in the estimate. As a result, great caution should be taken when determining the value of an intangible benefit in breakeven analysis. For additional information on breakeven methodologies for IT investment analysis see Sassone (1988) and Gallagher (1974).

A statistical approach to non-quantifiable benefits

One way to measure or estimate the value of *non-quantifiable benefits* is through the use of "regression analysis". *Regression analysis* is a statistical procedure that can take one or more independent variables, like the subjective values for turnover or quality, and compute a linear function estimate for a dependent variable, like costs. To avoid the computational aspects of regression analysis, we will be using Microsoft's© Excel© computer program for the computations. (For a basic review of this methodology reference Excel's HELP window on Statistic Functions, any basic statistics book, or see Meredith, *et al.* (2002, pp. 110-118).)

Let's illustrate how we can use this methodology to estimate costs in absenteeism. Assume a telemarketing company has a staff of over 200 employees that use a variety of telecommunication systems to contact customers to sell products and services. The company wants to buy a new software application that will more accurately tract employee bonuses and ear mark them for employees who more reliably come to work and produce for the company. The acquisition of the software it is felt will reduce absenteeism. Prior experience with other companies reveals at minimum of 25 percent reduction in absenteeism.

The company would like to estimate what absenteeism is costing them now so they can factor this into their "benefits" side of the investment analysis. To do this, the firm collected the data in Table 4. The observed relationship between the number of daily calls to customers (which means sales to the company) and the number of employees absent are obviously, inversely related. The loss in calls per day due to an absent employee can be estimated with the use of a regression model from Excel's© software system. In Excel©, the listing of Statistical Functions includes the SLOPE. This function takes the data and converts it into a simple regression linear model useful for estimating a linear trend in a set of data. The "slope" of the line given by Excel© will provide the estimate of the relationship between the two variables in this example. In this case entering the dependent variable of daily sales and the independent variable number of employees absent, the resulting linear regression slope value is -76.9. What this represents is the sales calls lost for each day an employee is absent. So in this example, every day an employee is absent, the company loses 76.9 calls to customers that might have generated sales for the company. It is common in any sales operation, particularly telemarketing, to be able to know the average dollar sales per call made, and so, a final value for the total cost of absenteeism can be determined. This analysis can be extended even further to estimate the error in the estimate by computing the *standard error of estimate* (see Lee *et al.*, 1998, pp.547-550), which is a standard deviation for the estimates of the independent variable. This statistic can be used to determine an interval estimate over which the parameter, in this case the loss of calls per day, will be accurate.

Table 4. Sample data for absenteeism example.

Days of the week	Dependent variable: Daily sales calls made to customers	Independent variable: Number of employees absent per day
Monday	11,100	12
Tuesday	11,375	7
Wednesday	11,160	16
Thursday	10,096	23
Friday	11,077	14

What is Payback Period Methodology?   → page 93

Payback period methodology is a traditional capital budgeting technique used to evaluate capital investments where the payback period of an investment is compared to some pre-specified length of time, referred to as the cutoff period. The *payback period* is the amount of time required to recover the cost of the initial investment. The *cutoff period* is the pre-specified length of time in which an investment must recover its initial investment to be considered the best alternative. The decision rule for the payback period methodology is fairly simple. If the payback period is shorter than or equal to the cutoff period, make the investment; if it is not, do not make the investment. When selecting one or more alternatives among a set, the investment or investments associated with payback periods less than or equal to the cutoff period are considered to be the best and should be undertaken.

Payback period methodology involves selecting a suitable cutoff period and determining the payback period for each alternative investment. Most firms and banking officials are aware of ideal benchmarks of payback periods. Each type of alternative investment in IT has its own ideal payback period. Equipment like PC's, for example, would have to have a very short payback period, usually two years, in order to justify the investment due to their relative speedy obsolescence time period.

Let's illustrate the use of the payback period methodology with an example. Suppose that an organization must select one computer system from a set of two alternative systems. Table 5 presents the initial investment and cash flows associated with the two alternative investments. Let's assume that it has been determined that the cutoff period is 2 years. The alternative IT investment that recovers the initial cost in 2 years or less is the best alternative according to this methodology and the specified parameter.

Table 5. Data for a payback period problem.

	Alternative A Computer System	Alternative B Computer System
Initial cost ($)	40,000	40,000
Cash flow year 1 ($)	30,000	20,000
Cash flow year 2 ($)	15,000	10,000
Cash flow year 3 ($)	25,000	30,000
Cash flow year 4 ($)	35,000	40,000

As shown in Table 6, the payback period for Alternative A Computer System is 2 years and that for Alternative B Computer System is 3 years. Accordingly, Alternative A Computer System is the best as it recovers the initial investment in 2 years.

In this example there is no question as to which alternative is the best and it appears that the best alternative is actually the best over the entire life-cycle of the investment, since its final cash flow value of $105,000 is greater than the alternative at only $100,000. However, many regard the payback period methodology as a simple rule of thumb rather than a reputable methodology because of its disadvantages. The first disadvantage is that the methodology ignores the time value of money.

As a result, cash flows received in the future are worth less than those received today, so those received in the future should be discounted to reflect the amount lost. The payback period methodology equally weighs all cash flows, irrespective of the time period they are to be received *Net present value* (NPV) is one method that considers the time value of money and the NPV of each alternative in the previous problem is shown Table 7. Alternative A Computer System is associated with the larger NPV and confirms that in this case the payback period methodology actually identifies the best investment. In other situations, as we will see in a subsequent problem, this is not always the case.

Table 6. Data for a payback period revised problem.

	Alternative A Computer System	Sum of cash flow for Computer A	Alternative B Computer System	Sum of cash flow for Computer B
Initial Cost ($)	40,000	-	40,000	-
Cash flow year 1 ($)	30, 000	30,000	20,000	20,000
Cash flow year 2 ($)	15,000	45,000	10,000	30,000
Cash flow year 3 ($)	25,000	70,000	30,000	60,000
Cash flow year 4 ($)	35,000	105,000	40,000	100,000

Table 7. Results for payback period problem.

	Initial investment ($)	Payback period	NPV @ 10% ($)
Alternative A Computer System	40,000	2 years	42,357.76
Alternative B Computer System	40,000	3 years	36,306.26

Another disadvantage of the payback period methodology is that cash flows beyond the payback period are not considered in the analysis.

IT investments that return little during the payback period and return a lot in the periods subsequent to the payback period are deemed inferior to those that recover the initial investment in the shortest amount of time. Again this is not a problem for the above example, but can be demonstrated.

Let's suppose that an organization must select one computer system from a set of two. The initial investment and cash flows are presented in Table 8. Assume that the cutoff period is two years.

Table 8. Data for alternative payback period example.

	Alternative A Computer System A	Sum of cash flow for Computer A System	Alternative B Computer System B	Sum of cash flow for Computer B System
Initial Cost ($)	100,000	-	100,000	-
Cash flow year 1 ($)	100,000	100,000	0	0
Cash flow year 2 ($)	5,000	105,000	50,000	50,000
Cash flow year 3 ($)	5,000	110,000	250,000	300,000

As shown in Table 9, according to the payback period methodology, Alternative A Computer System is the best alternative as the payback occurs within the first year. The initial investment of $100,000 is recouped in just 1 year for the Alternative A Computer System, while Alternative B Computer System requires 3 years to recoup the initial investment. However, when considering the time value of money and all cash flows, not just those within the cutoff period, Alternative B Computer System has the higher NPV and is the best alternative. NPV analysis is presented in Chapter 5 and is a discounted cash flow technique that considers the time value of money and all cash flows. Notice that the NPV associated with Alternative B Computer System, $141,325, is considerably larger that that of Alternative B Computer System, $848.45.

In situations like this one, the payback period methodology provides a conflicting solution to the NPV analysis. Some may argue that the payback period methodology is inferior and provides an incorrect result in this situation. Because the net present value analysis considers the time value of money and all cash flows, it yields the best solution. On the other hand, many IT investment decisions are not made based on longer-term cash flows, but how fast the investors can re-couple their investment. This conflict of personal objectives, which can impact investment methodology selection, is one that can only be resolved by the individual's preferences of the analyst and the goals stated for the investment decision they face.

Table 9. Results for a revised payback period problem.

	Initial investment ($)	Payback period	NPV @ 8% ($)
Alternative A Computer System	100,000	1 year	848.45
Alternative B Computer System	100,000	3 years	141,325.00

One way to combat the disadvantages of the payback period methodology is to select an appropriate cutoff period. If an organization uses the same short (two or three year) payback period, regardless of project life, more short-lived investments will be accepted than longer-lived ones. As a result, the organization will be focusing on the short-term and ignoring the long-term. By using longer-term cutoff periods when appropriate this problem can be lessened. Another way to avoid problems is to consider the time value of money—discount the cash flows before calculating the payback period of alternative IT investments. The *discounted payback period* is the number of periods necessary to recover the initial cost using the present value of the cash flows. In sum, discounting cash flows and using an appropriate cutoff period are ways to mitigate the disadvantages associated with the payback period methodology.

Despite the aforementioned shortcomings, the payback period methodology is more extensively used than the more advanced methodologies. According to a recent survey by Deloitte & Touche, out of 200 Chief Information Officers on how they measure the value of their IT investments, 54 percent said payback was used (IT Value, 2003). One of the reasons why it is so used is that the payback period methodology is easy to understand and apply. Evaluators can easily calculate the payback period and convey the results to decision-makers without having knowledge of the time value of money. Everyone in business can understand the need for a particular project recouping the initial investment in only two years is more favorable than one that requires five years. In addition, in the case of IT investment decision-making, the rapid change associated with information technology may cause obsolescence in only a few short years and thus, justify the use of the payback period methodology and short cutoff periods.

What is Accounting Rate of Return Methodology?

Accounting rate of return methodology is another technique typically used in capital budgeting decision-making, where the accounting rate of return is compared to a cutoff rate of return. *Accounting rate of return* (ARR), also called *book rate of return*, is the average annual income from an IT investment divided by average annual book value of the initial investment cost. ARR is like any typical return, measuring income or cash flows as a proportion of the initial investment amount. The cutoff rate of return is a hurdle rate set to determine the attractiveness of alternative investments. Just as with the cutoff period used in the payback period methodology, the cutoff rate of return must be carefully set and is used to determine if an investment should be undertaken and/or to determine the best investment. ARR methodology may be used to evaluate a single investment or to select one or more from a set of alternatives investments.

ARR is calculated as:

$$ARR = \frac{\text{average annual net income}}{\text{average annual book assets}}$$

Average annual net income and *average annual book assets* are calculated with values that would appear on the financial statements of an organization, and, thus are subject to accounting rules and regulations. These values are different than those utilized in other methodologies because other methodologies use estimated cash flows attributable to the IT investment versus straight accounting values.

Let's illustrate the use of this methodology with a simple example. Suppose an organization must decide whether or not to invest in a computer system. The initial cost is $90,000 and the investment is depreciated straight-line over three years. The accounting rate of return must be equal to or greater than the cutoff rate of return for the investment to be undertaken. Assume the cutoff rate of return is 10 percent. Both average annual net income and average annual book assets must be calculated to determine the accounting rate of return for the potential investment. Table 10 shows the accounting data necessary to determine the average annual income.

Table 10. Average annual net income data.

	Year 1	Year 2	Year 3
Cash inflow ($)	50,000	30,000	40,000
Depreciation ($)	30,000	30,000	30,000
Net income ($)	20,000	0	10,000

Average annual net income = $10,000

As shown in Table 10 the cash inflow less depreciation gives the net income for each year of the IT investment. An average is taken to provide an average annual net income of $10,000 for the computer system. Table 11 shows the accounting data necessary to determine the average annual book value of assets (for the IT investment).

Average annual book value of assets is calculated by first determining the net book value of assets, which is the gross book value of assets less cumulative depreciation. An average is taken of the net book value of assets for the four years, which yields an average annual book value of assets of $45,000. The accounting rate of return is 22.2 percent (10,000 / 45,000 = .222). The accounting rate of return is larger than the cutoff rate of 10 percent and, thus, the investment should be undertaken.

Table 11. Average annual book value of assets data.

	Year 0	Year 1	Year 2	Year 3
Gross book value of asset ($)	90,000	90,000	90,000	90,000
Cumulative Depreciation ($)	0	30,000	60,000	90,000
Net book value of asset ($)	90,000	60,000	30,000	0
Average annual book assets = $45,000				

Accounting rate of return may also be calculated as:

$$ARR = \frac{\text{average annual net income}}{\text{initial required investment}}$$

The difference from the previous calculation is that the initial cost of the investment is used instead of the annual book value of assets (see Horngren, 1981, for this variation). As a result, the accounting rate of return is calculated as the average annual net income divided by the initial cost of the investment. Calculated in this way using the same problem as above, the accounting rate of return is 11.1 percent (10,000 / 90,000 = .111). Again the organization should invest in the computer system because the accounting rate of return is larger than the cutoff rate of return given in the problem. The decision as to what term to use in the denominator for the accounting rate of return computation should be made by the decision maker, as both methods are used and presented in finance and accounting.

Accounting rate of return methodology, like the payback period methodology, is simple to use, present, and understand. However, the accounting rate of return does not consider the time value of money. In addition, it uses book values, which are dependent on the type of depreciation method utilized and the decisions concerning capitalization versus expense. In our example above, straight-line depreciation was used, but now let's consider the following accelerated depreciation charges of $50,000, $20,000 and $20,000 in years 1, 2, and 3, respectively. Table 12 and 13 show the revised calculation necessary for employing the accounting rate of return methodology. Table 11 showed that the average annual net income remains unchanged at $10,000. However, the average annual book assets does change from $45,000 to $37,500. The accounting rate of return is now 26.7 percent (10,000 / 37,500 = .267). For this particular problem, using accelerated depreciation instead of straight-line depreciation improves the accounting rate of return but does not change the solution afforded by the methodology.

Table 12. Revised average annual net income data.

	Year 1	Year 2	Year 3
Cash inflow	50,000	30,000	40,000
Depreciation	50,000	20,000	20,000
Net income	0	10,000	20,000

Average annual net income = 10,000

Table 13. Revised average annual book value of assets data.

	Year 0	Year 1	Year 2	Year 3
Gross book value of asset	90,000	90,000	90,000	90,000
Cumulative Depreciation	0	50,000	70,000	90,000
Net book value of asset	90,000	40,000	20,000	0

Average annual book assets = 37,500

Book values are also affected by the decisions accountants make as to whether a cost associated with an IT investment is an operating expense or part of the capital investment. Suppose an organization is purchasing a computer system and a service package. This computer system will require hardware, software, personnel, training, and maintenance. According to company standards, hardware, software, personnel, and vendor service are considered part of the capital investment, while training and maintenance are operating expenses. However, in other situations vendor service and possibly training may be considered operating expenses instead of capital expenditures. The classification of costs may greatly affect the accounting rate of return and thus the investment decision.

In general, if a cost is considered part of the capital investment instead of an operating expense, all else held constant, then average annual net income and average annual book assets should both increase; therefore, increasing the accounting rate of return. If the opposite occurs and a cost is considered an operating expense instead of a capital expenditure, then the accounting rate of return ratio should decrease. Analyzing and understanding the affects of cost classification can improve the understanding of the accounting rate of return and the corresponding decisions made according to its rules. A sensitivity analysis may be conducted to determine the effects of a cost that may be considered as either part of the capital expenditure or as an operating expense. If the cost classification affects the accounting rate of return enough to change the solution, then great care should be taken in actually classifying the cost. It may be that for IT investment evaluation purposes, the cost may be reclassified to best reflect the actual situation. Decision makers should be aware of cost classification and its affects on accounting rate of return methodology.

Summary

This chapter presented several IT investment techniques referred to as "basic financial methods." These investment techniques included breakeven analysis, payback period, and accounting rate of return. The limitations and computational procedures for these methodologies were discussed. Suggestions on how the limitations can be overcome and methodologies for computing intangibles were also presented.

There is a very large body of financial methodology that exists in the literature. A summary of this prior literature will be presented in the next chapter as a means of providing a context of the sheer volume of techniques available for IT investment decision-making. In the next chapter we will also explore the methodological procedures of "other financial methods" that are related to those presented in this chapter but helping to extend their capacities to deal with even more complex IT investment issues.

Review Terms

Accounting rate of return (ARR)
Accounting rate of return methodology
Average annual net income
Advanced financial methods
Annual cost
Average annual book assets
Basic financial methods
Book rate of return
Breakeven analysis
Capital assets
Capital budgeting decisions
Cutoff period

Discounted payback period
Initial cost
Net costs
Net present value (NPV)
Non-quantifiable benefits
Payback period
Payback period methodology
Present value (PV)
Real assets
Regression analysis
Standard error of estimate

Discussion Questions

1. Why use "breakeven analysis" in IT investment decision-making?
2. Why are non-quantifiable benefits an issue in "breakeven analysis"?
3. Why use the "payback period method" in IT investment decision-making?
4. Why use the "accounting rate of return" in IT investment decision-making?
5. Why use "breakeven analysis" over say the "payback period method"?

6. Explain how not taking into consideration the time value of an investment can cause a problem when using the "payback period method"?

7. Why is the classification of costs so important when using the "accounting rate of return" in IT investment decision-making?

8. Why use "payback period method" over say the "accounting rate of return"?

9. Why were two formulas for "accounting rate of return" given? Explain the need for both.

10. How can accounting book values disrupt an "accounting rate of return" analysis?

Concept Questions

1. What are the steps and the information needed in order to conduct a "breakeven analysis"?

2. What are some of the limitations and disadvantages in using "breakeven analysis"?

3. How are "present values" used in "breakeven analysis"?

4. What is the difference between a "real asset" and a "capital asset"?

5. How can you use "regression analysis" to estimate non-quantifiable benefits?

6. What are the steps and the information needed in order to use "payback period method"?

7. What is a "cutoff period" used for in "payback period method"?

8. Can you make a decision using the "payback period method" without a cutoff period?

9. What are some of the limitations and disadvantages in using the "payback period method"?

10. How is "net present value" used in the "payback period method"?

11. What are the steps and the information needed in order to compute an "accounting rate of return"?

12. What are some of the limitations and disadvantages in using an "accounting rate of return"?

Problems

1. A computer firm has just conducted a costs and benefits analysis for a breakeven study they are performing on the purchase of an upgrade to their central processing unit (CPU) technology. The initial cost of the technology is $250,000, and new training for tech personnel will run $20,000 for the first four years. The improved speed of their supply-chain computer will automatically qualify the firm for new business contracts. These contracts will mean additional cash flow benefits of $100,000 per year for five years. Without considering the time-value of costs or benefits, is this a good investment based on breakeven analysis logic?

2. (Refer to Problem 1) Given the stated costs, the benefits change, and are now estimated to be the following during the five-year life of the CPU: Year 1 = $15,000, Year 2 = $50,000, Year 3 = $100,000, Year 4 = $85,000, and Year 5 = $30,000. Without considering the time-value of costs or benefits, is this a good investment based on breakeven analysis logic?

3. The amount of data storage companies are facing today in light of data-intensive graphics, Web transactions, and other digital applications, requires most firms to double their storage needs every 12 to 18 months. One IT called Redundant Array of Inexpensive Disks (RAID) is now meeting that challenge by packaging over a hundred disk drives, a controller chip and specialized software into a single large storage unit. A firm is planning on introducing a RAID system that will cost $500,000 and have yearly costs and estimated cash flow benefits as stated in the table below. While the benefits of increased sales were forecast based on contractual opportunities the RAID system would bring to the firm, the productivity was based on the fact that one data entry employee could be terminated and the yearly salary of $30,000 saved. Since the loss of the person might negatively impact employee morale, the costs of this negative value were not included as a benefit. Without considering the time-value of costs or benefits, is this a good investment based on breakeven analysis logic?

	Year 0	Year 1	Year 2	Year 3	Year 4
Costs:					
Initial cost	500,000				
Annual cost		25,000	50,000	75,000	100,000
Benefits:					
Increased sales		120,000	150,000	150,000	180,000
Improved productivity		30,000	30,000	30,000	30,000
Improved employee morale		--	--	--	--

4. We are looking at choosing one of two investments: A or B. Investment A has a PV of costs at $145,000 and a PV of benefits at $150,500. Investment B has a PV of costs at $275,000 and a PV of benefits at $300,000. If you can only choose one investment, which would be it? Explain your answer.

5. We want to invest in one of two servers (either server A or B). Each has advantages and differing costs to the firm for the purchase and implementation of the software. The initial costs of the technology for the firm are given in the table below. The estimated cash flow generated as a direct result of the servers being available to the firm's customers is also presented for both servers in the table below. Based on payback period method, where the shortest payback period determines the best choice, which server should be selected?

	Server A	Server B
Initial Cost ($)	110,000	140,000
Cash flow year 1 ($)	30, 000	40,000
Cash flow year 2 ($)	50,000	60,000
Cash flow year 3 ($)	25,000	60,000
Cash flow year 4 ($)	40,000	40,000

6. (Refer to Problem 5). If we add a yearly service contract costing of $10,000 per year, to the cost-side of the problem, which of the two servers would you recommend as an investment alternative? Does it change your solution if you have a cutoff period of three years?

7. An e-retailer is losing a lot of sales because of their inability to provide their online customers with answers to questions when

they fill out order forms. The e-retailer has decided to purchase the best query language software available in the market. They want to invest in one of three query languages (either A, B, or C). Each software has advantages and differing costs to the firm for their purchase and implementation. The initial costs of the software to the firm are given in the table below. The estimated cash flow generated as a direct result of the software saving lost sales and being available to the firm's customers is also presented in the table below. Based on payback period method, where the shortest payback period determines the best choice, which query language software should be selected?

	Software A	Software B	Software C
Initial Cost ($)	50,000	65,000	70,000
Cash flow year 1 ($)	10, 000	20,000	30, 000
Cash flow year 2 ($)	15,000	10,000	45,000
Cash flow year 3 ($)	25,000	30,000	25,000
Cash flow year 4 ($)	35,000	40,000	35,000

8. Storage of data for a small company has become a problem. The company is trying to decide which of two storage service providers (SSP) they should subscribe to. An SSP is a third party provider that rents out storage space to subscribers over the Web. The SSPs require an initial investment to install their Web communication software on the subscriber company's mainframe system. The initial investment, payback period and net present value of the future cash savings (over purchasing and owning additional IT) is given in the table below. Based on the information in the table below, which SSP should be selected? Explain your answer.

Alternative storage service providers	Initial investment ($)	Payback period	NPV @ 8% ($)
SSP A	20,000	2 years	35,589.21
SSP B	15,000	3 years	32,888.87

9. Based on the information in the table below, which alternative should be selected? Defend your answer regardless of what it maybe. Hint: There maybe more than one correct answer to this problem.

Alternatives	Initial investment ($)	Payback period	Life of investment	NPV @ 12% ($)
A	200,000	2 years	4 years	699,891
B	210,000	3 years	3 years	700,456
C	175,000	5 years	5 years	902,411
D	100,000	4 years	4 years	588,991
E	121,000	1 year	2 years	140,000

10. A company has determined their average annual net income from an IT investment in computers to be $125,000 and their average annual book asset value for that IT investment is $1.5 million. What is their ARR? What does this proportion represent? Explain.

11. In the past, a company used an ARR of 10 percent as a guideline to IT investment purchases. Based on that decision criterion three years ago the company invested in a mini computer system. The three years since its purchase, the company has maintained information on its contribution to cash flow and other accounting information as presented in the two tables below. They now want to post-evaluate that IT investment decision. What is the computed ARR on this investment? Was it a good investment relative to their guideline of 10 percent?

	Year 1	Year 2	Year 3
Cash inflow ($)	60,000	60,000	60,000
Depreciation ($)	40,000	40,000	40,000
Net income ($)	20,000	20,000	20,000

Average annual net income = $20,000

	Year 0	Year 1	Year 2	Year 3
Gross book value of asset ($)	120,000	120,000	120,000	120,000
Cumulative depreciation ($)	0	40,000	80,000	120,000
Net book value of asset ($)	80,000	80,000	40,000	0

Average annual book assets = $60,000

12. (Refer to Problem 11) Will changing the depreciation to an accelerated rate of $60,000, $40,000 and $20,000 respectively, for the three years, alter the ARR or the solution? Explain your answer.

References

Bacon, J.C., "The Use of Decision Criteria in Selecting Information Systems/Technology Investments," *MIS Quarterly*, September 1992, pp. 335-353.

Ballantine, J.A. and Stray, S., "Information Systems and Other Capital Investments: Evaluation Practices Compared," *Logistics Information Management*, Vol. 12, No. 1/2, 1999, pp. 78-   93.

Ballantine, J.A. and Stray, S., "Financial Appraisal and the IT Investment Decision-Making Process," *Journal of Information Technology*, Vol. 13, 1998, pp. 3-17.

Gallagher, Charles A., "Perceptions of Value of a Management Information System," *Academy of Management Journal*, Vol. 17, No. 1, March 1974, pp. 213-232.

IT Value. *The Controller's Report*, Vol. 2, No. 7, 2003, p. 7.

Lee, S.M., Schniederjans, M.J. and Olson, D.L., *Business Statistics: Quality Information for Decision Analysis*, Houston, TX: Dame Publications, Inc., 1998.

Meredith, J., Shafer, S. and Turban, E., *Quantitative Business Modeling*, Mason, OH: South-Western/Thomson Learning, 2002.

Sangster, A., "Capital Investment Appraisal Techniques: A Survey of Current Usage," *Journal of Business Finance and Accounting*, Vol. 20, No. 3, 1993, pp. 307-332.

Sassone, P.G., "A Survey of Cost-Benefit Methodologies for Information Systems," *Project Appraisal*, June 1988, pp. 73-84.

Tam, Yan Kar "Capital Budgeting in Information Systems Development," *Information & Management*, Vol. 23, 1992, pp. 345-357

Chapter 5

Other Financial Methodologies

Learning Objectives

After completing this chapter, you should be able to:

- Better understand the totality and variety of differing methods of information technology (IT) investment decision-making.
- Explain the difference between "ex ante" and "ex post" IT investment decision-making.
- Utilize "present value analysis" in IT investment decision-making.
- Explain what unequal investment lives can mean to an IT investment decision.
- Utilize "return on investment" methodology in IT investment decision-making.
- Utilize "internal rate of return" methodology in IT investment decision-making.

Introduction

In this chapter we will attempt to present a summary overview of the many financial and other methodologies useful in IT investment decision-making. The purpose here is to provide a basic listing of a variety of methodologies that have been used in IT investment decision-making. This chapter also provides a discussion and illustration of the procedural uses of several of the most common financial tools for evaluation IT investments. These include "present value analysis", "return on investment" methodology, and "internal rate of return" methodology. These methodologies and their computational procedures will be illustrated with examples.

What Types of Methodologies Support IT Investment Decision-Making?

For the last four decades, researchers and practitioners have been proposing methods to facilitate the IT investment decision-making process. Over fifty methods or techniques originating from the disciplines of accounting, management, and finance have been recommended by researchers and practioners (see Chan, 2000; Farbey *et al.*, 1994; Powell, 1992; Renkema and Berghout, 1997). These techniques range from traditional financial methods, like "payback period" and "net present value", to complex multi-criteria methods that incorporate both financial and non-financial criteria, like analytical hierarchy process and balanced scorecard. Before discussing other financial methodologies, a brief taxonomy of these methods will be presented that helps to categorize them based on the nature of their origin.

The various IT investment methods have been divided into the following groups or categories: (1) Financial Techniques; (2) Operations Research/Management Science Techniques; (3) Techniques Specifically Designed for IT Evaluation; and (4) Other Techniques for IT Evaluation. The various techniques that fall into these four categories have been presented in the literature and/or are used in practice for the purposes of IT investment evaluation. The following discussion will provide a brief explanation of the most widely utilized methods within each category. Some of these methods can be characterized as *ex ante* methodologies (i.e., used prior to the IT decision as a means to compare alternatives prior to a choice) and/or *ex post* (i.e., used after the IT decision as a means of post evaluation to see if the IT investment met with desired success).

Financial Techniques for IT Investment Decision-Making

It has been argued that for the most part, financial techniques are not the optimal decision-making tools for IT investment decision-making because they fail to consider intangible costs and benefits. In some instances, IT investment costs (i.e., usually an example of *tangible* or *objective criterion*) and benefits are actually *intangible* or *subjective criteria*, like increased customer satisfaction, improved decision quality, and increased differentiation in products that may not be directly reflected in cash inflows from an IT investment. Financial techniques used in IT investment decision-making are rooted in finance and accounting. Table 1 shows the financial techniques and provides a description of each, as well as the type of criteria that the methodology utilizes and when the method is most likely to be used. Payback period, present value analysis, internal rate of return, accounting rate of return, and return on investment are commonly used financially based methodologies to evaluate information technology investments. These are traditional techniques that have been utilized by organizations for a number of years to evaluate capital investments. Because of their familiarity, these techniques are often the most widely used evaluation methods for technology investments. Other financial techniques have roots in accounting and include breakeven analysis, cost benefit analysis, and cost revenue analysis.

Payback period is a simple technique in which the time period necessary to recoup the initial investment is calculated and used to evaluate an investment and/or a set of mutually exclusive investments. *Net present value* and *internal rate of return* are based on a well-known corporate finance principle referred to as the time value of money. The principle states that the longer a return is deferred into the future, the lower its current value. So returns that will be realized further into the future are worth less than those realized sooner. As a result, cash inflows from a technology investment must be discounted and the present value of the investment is used to evaluate whether or not to invest. *Accounting rate of return* and *return on investment* are simple calculations that provide the decision maker with a ratio to evaluate the investment decision. The ratio of expected profit to initial investment cost is compared to the opportunity cost of capital; if the return is greater than the opportunity cost of capital the investment should be undertaken.

Cost revenue analysis involves comparing the costs with the benefits of the investment that can be directly attributed to the computer system. Benefits that can be directly attributed to the technology investment usually include some type of cost savings the technology provides. *Cost benefit analysis* is an extension of cost revenue analysis where the costs of the technology are compared to the benefits that can be directly and indirectly attributed to the system. *Breakeven analysis* can be used in many ways to evaluate IT investments, however, it is most often used by comparing the present value of the costs with the present value of the benefits of the investment. A more in depth discussion of financial techniques for IT investment decision-making may be found in this chapter and others in this textbook.

Operations research/management science techniques for IT investment decision-making

Unlike many financial IT methodologies, *operations research/ management science* (OR/MS) are methods that are based on mathematics, engineering, algorithms, heuristics and other methods from the field of applied mathematics called OR/MS. They collectively represent a set of techniques that may be used to incorporate intangible or subjective criteria directly into the decision-making process. As with most types of decision-making, decision makers are faced with a problem and must make a decision to solve that problem. There may be several decision alternatives each with consequences and OR/MS techniques have been developed to assist decision makers choose, as rationally as possible, the best alternative.

Table 1. Financial techniques for IT investment decision-making.

	Description	Type of criteria	Evaluation timing
Accounting rate of return	Compare the average after-tax profits with initial investment cost	Tangible	Ex ante or ex post
Breakeven analysis	Compare the present value of costs with the present value of benefits	Tangible and intangible	Ex ante, most often
Cost benefit analysis	Compare costs with benefits that can be directly and indirectly attributed to the system	Tangible and intangible	Ex ante or ex post
Cost benefit ratio	Calculate the ratio of costs of an IT investment to its benefits measured in monetary terms and compare to a threshold ratio	Tangible	Ex ante and ex post
Cost revenue analysis	Compare costs with benefits that can be directly attributed to the system	Tangible	Ex ante, most often
Internal rate of return	Calculate the return that equates the net present value of an investment to zero	Tangible	Ex ante or ex post
Net present value analysis	Discount cash inflows and compare them to cash outflows	Tangible	Ex ante or ex post
Payback period	Calculate the time required to recoup initial cost	Tangible	Ex ante, most often
Profitability index	Calculate the per dollar contribution of an investment	Tangible	Ex ante or ex post
Return on investment	Calculate the return of an investment	Tangible	Ex ante or ex post

Table 2 shows that analytic hierarchy process, decision/Bayesian analysis, Delphi evidence, game playing, multi-objective, multi-criteria approaches and simulation have been and are currently used for IT investment evaluation and selection. These methods are widely known and used OR/MS techniques. It should be noted that there are a number of multi-objective, multi-criteria approaches, including but not limited to, mathematical programming, factor-rating methods and experimental methods that have been applied in IT investment decision-making.

Table 2. Operations research/management science techniques for IT investment decision-making.

	Description	Type of criteria	Evaluation timing
Analytical hierarchy process	Calculate the score of decision-makers' pair wise comparisons	Tangible and intangible	Ex ante
Decision/Bayesian analysis	Calculate the expected value of investing in alternative investments	Tangible and intangible	Ex ante
Delphi evidence	Obtain consensus of experts' opinion concerning the best alternative investment	Tangible and intangible	Ex ante and ex post
Game playing	Calculate payoff of investment based on actions of the competition, mathematics, and economic theory	Tangible and intangible	Ex ante
Multi-objective, multi-criteria approaches	In general, develop a measure of utility provided by an IT investment	Tangible and intangible	Ex ante and ex post
Simulation	Model how an investment will perform and impact the organization	Tangible and intangible	Ex ante and ex post

Techniques specifically designed for IT investment decision-making

Both practitioners and researchers have designed and utilized a myriad of techniques, specifically for use in IT investment decision-making. Due to the unique nature of IT investments, many contend that the techniques used to evaluate and select the IT should fit the particular type of IT investment and the particular organizational management style. Many of the twenty-plus methods listed in Table 3 have been developed for a specific type of technology and/or for a particular organization, thus creating a fit between the type of technology and the investment methodology and/or a "fit" (note Chapter 1 about organization fit) between the organization and the type of investment methodology. In addition, some methods have been developed to complement the more traditional financial methods and may be used as supplemental methods to reflect the intangibles qualities of IT.

Information economics and return on management have received much coverage in the literature. *Information economics*, developed by Parker *et al.* (1989), is an all-encompassing technique that involves the analysis of qualitative as well as quantitative factors and considers business and technology risk *Return on management* is another popular technique based on the assumption that IT primarily improves the productivity of management. Return on management is a productivity ratio of the value-added by management due to IT compared to the total cost of management. The drawback of return on management is that the value-added by management is a residual value and thus the method may be best used in conjunction with other methods for ex post evaluation.

Portfolio management principles and procedures may be successfully applied to IT investment decision-making. Large organizations may have hundreds of different IT investment projects underway at one time. These IT projects may be collected and categorized in a portfolio of projects. Just as one manages a portfolio of stocks, a portfolio of IT investments may be used to monitor existing IT investments and evaluate new ones by cost, benefit, and risk. Bedell's method, investment mapping, investment portfolio and Ward's portfolio approach are portfolio approaches specifically designed for IT investment decision-making.

Several methods have been developed and used by IBM for the evaluation and selection of IT technology. These methods assist decision makers by providing a means to value an IT investment and include SESAME, systems investment methodology, strategic application search, application transfer team, automatic value points, executive planning for data processing, information system investment strategies, and process quality management.

Table 3. Techniques specifically designed for IT investment decision-making.

Technique	Description	Type of criteria	Evaluation timing
Application benchmark technique	Construct a computer program to be run by vendors so as to determine the run time of individual computer system configurations	Tangible	Ex ante
Application transfer team	Conduct a study to determine exact requirements of the IT and to support the business case	Tangible and intangible	Ex ante
Automatic value points	Calculate the degree of automation based on a set of criteria concerning the contribution of IT to the overall business performance	Tangible and intangible	Ex ante and ex post
Bedell's method	Calculate contribution of an IT system by multiplying an importance score by the level of quality improvement made by the system	Tangible and intangible	Ex post and ex ante
Benefit/Risk analysis	Determine the overall risk of an investment through the use of a risk questionnaire and judge whether the benefits outweigh the risks	Intangible	Ex ante
Buss's method	Determine investment priority by ranking alternative investments and those with the highest frequency have highest priority	Tangible and intangible	Ex ante
Cost-value technique	Total all costs associated with a system and deduct an earned value (established dollar value of desirable feature minus vendor charge for desirable feature)	Tangible	Ex ante

Table 3. Techniques (Continued).

Technique	Description	Type of criteria	Evaluation timing
Executive planning for data /processing	Conduct a comparison of costs and benefits of existing systems and examine areas for future investment	Tangible and intangible	Ex ante
IT assessment	Calculate financial and non-financial ratios and compare them to benchmark ratios	Tangible and intangible	Ex ante and ex post
Information economics	Calculate the overall value of an investment based on enhanced ROI, business domain, and technology domain criteria	Tangible and intangible	Ex ante
Information systems investment strategies	Make a financial comparison between the organization and its competitors, examine the portfolio of existing applications and prepare the business case for areas with expected high returns	Tangible and intangible	Ex ante
Investment mapping	Calculate evaluation criteria scores and plot investment alternatives on a grid	Tangible and intangible	Ex ante and ex post
Investment portfolio	Calculate contribution of IT system to business and technology domain and calculate financial consequences (NPV) of the system	Tangible and intangible	Ex ante
Knowledge based system for IS evaluation	Obtain an overall quantitative rank based on traditional capital budgeting techniques and an overall qualitative rank of projects based on rules established by MIS planning groups and MCDM models	Tangible and intangible	Ex ante
MIS utilization technique	Calculate the overall success of an IT investment based on 48 performance criteria	Tangible and intangible	Ex post
Process quality management	Analyze mission, critical success factors and key business processes to identify areas for IT investment	Tangible and intangible	Ex post

Table 3. Techniques (Continued).

Technique	Description	Type of criteria	Evaluation timing
Requirements-costing technique	Calculate total cost of an investment as cost of the mandatory features plus additional costs for desirable but not included features	Tangible	Ex ante (mostly for vendor selection decisions)
Return on management	Calculate the return of an investment that can be attributed to management productivity	Tangible	Ex post, most often
SESAME	Compare the cost of a computer system with the cost of performance without a computer system	Tangible	Ex ante and ex post
SIESTA	Assess benefits and risks of the fit between IT technology strategy/infrastructure and business strategy/infrastructure	Mostly intangible	Ex ante and ex post
Strategic application search and systems investment methodology	Analyze the extent of existing systems and identify the most productive areas for future investment	Intangible	Ex ante, most often
Value analysis	Establish value of a system (and/or prototype) by asking management simple value-related questions and compare that value to investment cost	Tangible and intangible	Ex ante
Ward's portfolio approach	Assess risk of investment and risk of the portfolio of investments after undertaking investment	Tangible and intangible	Ex ante and ex post
Zero-based budgeting	Partition projects into smaller projects, assess each smaller project based on the same evaluation framework, and select the most important smaller projects assuming limited funding	Tangible and intangible	Ex ante

Other techniques for IT investment decision-making

Table 4 shows other techniques that have been put forth in the literature and are used in practice for IT investment decision-making. Many of these techniques are frequently used in management decision-making and have been adapted for use in IT evaluation and selection. The *balanced scorecard* is a popular technique used to evaluate organizational performance developed by Kaplan and Norton (1992) and was adapted for use in IT evaluation and selection by Douglas and Walsh (1992).

Table 4. Other techniques for IT investment decision-making.

Technique	Description	Type of criteria	Evaluation timing
Balanced scorecard	Evaluate an investment from the user's, business value, efficiency, and innovation/ learning perspectives	Tangible and intangible	Ex ante and ex post
Boundary values/spending ratios	Calculate the ratio of IT cost to a known aggregate value (total sales, total assets, etc.)	Tangible	Ex post, most often
Cost displacement/ avoidance	Compare the cost of IT investment to the current costs displaced by the IT system plus the projected costs avoided by the system	Tangible	Ex ante
Cost effectiveness analysis	Compare the effectiveness of a system with its cost and select the system with the lowest cost, best effectiveness, or the optimal combination of both	Tangible and intangible	Ex ante and ex post
Critical success factors	Obtain, compare, and rank factors critical to business success, and based on these rankings, deduce investment priorities	Intangible	Ex ante

Table 4. Other techniques (Continued).

Technique	Description	Type of criteria	Evaluation timing
Hedonic wage	Based on employee activity time allocation, calculate the marginal value of each employee and use these values to estimate the value of IT investment benefits	Tangible	Ex ante and ex post
Real options valuation	Calculate the additional value of an investment that exists because it provides the option for a second investment	Tangible and intangible	Ex ante
Quality engineering	Translate perceived value and risk into a quality score	Intangible	Ex ante and ex post
Satisfaction/priority surveys	Survey and compare user and IS professionals' opinions on the effectiveness and importance of installed systems	Intangible	Ex ante and ex post
Structural models	Create a model to analyze how an information system affects the costs and revenues of the particular business function or line of business it is intended to serve	Tangible and intangible	Ex ante and ex post
Time savings times salary	Calculate the value added of an IT investment by estimating the percentage of time the system will save workers and multiply by the cost of the workers	Tangible	Ex ante
Value chain analysis	Assess how an IT investment can provide competitive advantage in each phase of the chain	Tangible and intangible	Ex ante and ex post

Real options valuation is another technique that has received much attention for IT investment decision-making. Investing in one type of IT may provide the option of investing in another type; thus the additional value of this option should be considered in the evaluation of the original investment. Real options valuation may be most appropriately used if there is high uncertainty in markets and the organization needs to stay flexible. It is usually used in conjunction with other methods for IT investment decision-making.

Cost displacement/avoidance, cost effectiveness analysis, hedonic wage, structural models and *time savings times salary* are cost justification techniques primarily based on economic theory and models. These cost displacement techniques are most useful for office-automation technology investments.

It should be noted that these listings of IT investment techniques are not exhaustive listings; a limited number of techniques have been proposed and/or are used for IT investment decision-making but were not included in this categorization because they are not utilized, are outdated and/or observed infrequently in the literature.

What is Present Value Analysis Methodology?

Present value analysis methodology is a traditional capital budgeting technique in which today's value of future cash flows of an investment are compared to the cost of the investment. According to a recent survey by Deloitte & Touche, out of 200 Chief Information Officers on how they measure the value of their IT investments, 29 percent said discounted cash flow were used (IT Value, 2003).

Today's value of future cash flows is referred to as the *present value* (PV) of the investment. In general, if the present value of an investment is greater that the cost of the investment then it should be undertaken because it will add value to the organization. Present value analysis is based on the basic assumption that a dollar today is worth more than receiving a dollar tomorrow because it can be invested and begin accruing interest immediately. Typically, present value analysis is used in situations where cash flows may be easily determined which tends to be when cash flows are due to a cost reduction or cost avoidance. Present value analysis may be used to evaluate an independent investment individually or to select among a set of mutually exclusive investments. Present value is calculated as:

$$PV = \frac{C_1}{1+r} + \frac{C_2}{(1+r)^2} + ... + \frac{C_n}{(1+r)^n}$$

where $C_1...C_n$ are the expected cash flows for n time periods, and r is the "discount rate." The *discount rate*, also called the *opportunity cost of capital*, is the rate that could be earned by investing in securities of comparable risk to that of the investment. The discount rate may be thought of as the expected return forgone by investing in the technology rather than in an equally risky investment in the capital market. An accurate estimate of the discount rate is necessary because this rate affects acceptance and rejection of individual investments.

A Payroll IT Investment Problem

To illustrate this methodology, lets look at an example. Suppose you are charged with the task of deciding whether or not to invest in a payroll system. A payroll system is a transaction processing system which will likely lead to a cost reduction for an organization. A new payroll system will cost $100,000 and the expected cost savings will be $40,000 per time period for the next four time periods. Financial management recommends a discount rate of 10% for this project. Should the organization invest?

Present value is today's value of future cash flows and the calculation for this payroll system is as follows:

$$PV = \frac{40,000}{1+.10} + \frac{40,000}{(1+.10)^2} + \frac{40,000}{(1+.10)^3} + \frac{40,000}{(1+.10)^4} = 126,794.62$$

The present value of the investment is \$126,893.94, which is larger than the initial cost of the investment of \$100,000. The investment is worth more to the organization than it costs so the organization should purchase the payroll system (assuming that the organization has enough funds to invest in the endeavor).

Net present value analysis methodology

Net present value (NPV) is another way of carrying out present value analysis. NPV is the present value of cash flows minus the initial investment cost and may be calculated as:

$$NPV = C_0 + \frac{C_1}{1+r} + \frac{C_2}{(1+r)^2} .... + \frac{C_n}{(1+r)^n}$$

where C_o is the initial investment, $C_1...C_n$ are the expected cash flows, r is the discount rate and n in the number of time periods. Investments with positive NPV should be undertaken because they add value to the firm. Given the above payroll IT investment problem NPV is:

$$NPV = -100,000 + \frac{40,000}{1+.10} + \frac{40,000}{(1+.10)^2} + \frac{40,000}{(1+.10)^3} + \frac{40,000}{(1+.10)^4} = 26,794.62$$

The organization should invest in the payroll system because it adds \$26,893.94 to the value of the company. It should be noted that the initial investment, C_o, is a cash outlay for the payroll system and is thus a negative number in the NPV equation. For an independent investment, the following rules may be used to guide the investment decision:

If NPV is greater than zero, then make the investment.
If NPV is less than or equal to zero, then do not make the investment.

The discussion thus far has focused on using present value analysis to evaluate an independent investment. Present value analysis may also be used to evaluate a set of mutually exclusive projects. Let's assume you are charged with the task of selecting among three alternative payroll systems. The costs of each system as well as the expected cash flows are presented in Table 5.

Table 5. Data for a present value analysis problem for mutually exclusive investments.

	Alternative A payroll system	Alternative B payroll system	Alternative C payroll system
Initial cost	$100,000	$100,000	$100,000
Cash flow year 1	$40,000	$25,000	$65,000
Cash flow year 2	$40,000	$30,000	$65,000
Cash flow year 3	$40,000	$45,000	$20,000
Cash flow year 4	$40,000	$55,000	$20,000

The NPV must be calculated for each alternative investment and then the alternative with the largest positive NPV should be selected to maximize the value of the organization. Table 6 summarizes the NPV calculations for each alternative utilizing a discount rate of 10 percent.

Table 6. Data for a present value analysis problem with mutually exclusive investments.

Alternatives	NPV calculation
Alternative A payroll system	$NPV = -100,000 + \dfrac{40,000}{1+.10} + \dfrac{40,000}{(1+.10)^2} + \dfrac{40,000}{(1+.10)^3} + \dfrac{40,000}{(1+.10)^4} = 26,794.62$
Alternative B payroll system	$NPV = -100,000 + \dfrac{25,000}{1+.10} + \dfrac{30,000}{(1+.10)^2} + \dfrac{45,000}{(1+.10)^3} + \dfrac{55,000}{(1+.10)^4} = 23,158.62$
Alternative C payroll system	$NPV = -100,000 + \dfrac{65,000}{1+.10} + \dfrac{65,000}{(1+.10)^2} + \dfrac{20,000}{(1+.10)^3} + \dfrac{20,000}{(1+.10)^4} = 41,546.14$

From this analysis, it can be seen that Alternative C Payroll System has the largest NPV and thus should be selected among the set of alternatives. Two things should be noted with respect to this set of alternatives. First, each alternative has an equal initial investment of $100,000 each. Under situations where the alternatives have unequal initial investments a "profitability index" may be calculated and used as the criteria for selection, especially when there are limited funds to make investments in computer systems. Second, each investment alternative has equal length or duration (i.e., four years). Not all investment alternatives in a set will have equal lives, and in these situations, one may choose to use one of several techniques to determine which alternative should be undertaken. The next two sections of this chapter address these issues.

Advantages of present value and net present value methodologies include the consideration of the time value of money and all cash flows. Disadvantages of PV and NPV can include the need to estimate opportunity cost of capital, the need to adjust for unequal lives and investment size, and the need for expertise for proper usage.

Computer-based solutions

A computer-based solution for net present value problems is available by using Microsoft© Excel©. By accessing Excel© and clicking the "Paste Function" key, a listing of "Function Categories" is provided. Clicking the "Financial" category reveals a "Function Name" listing, one of which is "NPV" or net present value. Clicking on this function reveals a window with open boxes that allow for the interest rate and as many payments as is desired in the problem. Future payments are expressed as negative values for capital investments and future income flows as positive values. In the case of Alternative A in Table 5, the values of interest at 0.10 and four years of $40,000 can be entered into the Excel© program. The entered values would be 40000, 40000, 40000, and 40000. The output is the future stream of discounted income, or $126, 794.62. Adjusting it for the $100,000 initial payment by manually subtracting it from the discounted income, we have as expected a NPV of $26,794.62.

Unequal investment sizes

Profitability index (PI) is a ratio that can be used to rank projects when the size of the initial investment varies for the alternative investments in a mutually exclusive set. PI is the ratio of NPV to the cost of the initial investment or:

$$PI = \frac{NPV}{Investment\ Cost}$$

Suppose that an organization is considering three payroll systems each with a different initial cost but an equal length of life. The cash flows and other essential data are presented in Table 7.

Table 7. Data for a present value analysis problem with unequal investment sizes.

	Alternative A Payroll System	Alternative B Payroll System	Alternative C Payroll System
Initial cost	$100,000	$10,000	$150,000
Cash flow Year 1	$40,000	$5,000	$70,000
Cash flow Year 2	$40,000	$5,000	$70,000
Cash flow Year 3	$40,000	$5,000	$10,000
Cash flow Year 4	$40,000	$5,000	$10,000

Table 8 shows the NPV and PI for each alternative system using a 10 percent discount rate. In terms of the contribution per dollar spent for each alternative system, Payroll System B is the best. Payroll System A and C only return $0.27 and $0.05, per dollar spent, while Payroll System B returns $0.58 per dollar. It should be cautioned that in situations with an unlimited or large amount of funds, NPV methodology may provide a better solution than PI.

Table 8. Profitability index for a present value analysis problem with unequal investment sizes.

	Initial investment	NPV	PI
Alternative A payroll system	100,000	26,794.62	26,794.62/100,000 = .27
Alternative B payroll system	10,000	5,849.33	5,849.33/10,000 = .58
Alternative C payroll system	150,000	7,345.81	7,345.81/150,000 = .05

Assuming the organization has unlimited funds, Payroll System A adds the most value ($26,795) and is the best alternative. When funds are limited, a situation described as being under capital rationing constraints, then Payroll System B provides the largest contribution per dollar spent, i.e., the biggest bang for the buck. Oftentimes organizations have a limited set of funds that may be allocated to investing in a particular type of IT technology. In these situations, investments with the highest PI will be selected until the limited amount of funds is exhausted. In general, PI may be best suited for use when capital is rationed and/or NPVs of the different investment alternatives are close and the difference between initial cost for the alternative investments is large.

Unequal investment lives

Several techniques exist to account for the problem of unequal lives of alternative investments. The simplest and most intuitive way is to adjust the project lives so that they are of the same length and then use NPV to evaluate the alternatives. The common denominator of the investments' lives is selected and the investments are extended to this length. To illustrate, suppose there are two alternative systems each costing $50,000; one has a useful life of two years and the other four years. Assume the discount rate is 10 percent and the cash flows for each alternative are those presented in Table 9.

Table 9. Data for a present value analysis problem with unequal investment lives.

	Alternative A payroll system	Alternative B payroll system
Initial cost	$50,000	$50,000
Cash flow Year 1	$40,000	$25,000
Cash flow Year 2	$40,000	$25,000
Cash flow Year 3	0	$25,000
Cash flow Year 4	0	$25,000

If we assume that Payroll System A is replicable (i.e., that another identical system can be purchased at the end of year two for $50,000), we can extend the life of Payroll System A to be equal to that of Payroll System B. The revised cash flows for Payroll System A for years one through four are $40,000, $-10,000, $40,000 and $40,000, respectively. Now that the projects are of equal length, the NPV is calculated and a decision is made based on this criteria. Table 10 shows that if Payroll System A is replicable, it is the best alternative.

Table 10. NPV replicated to a common denominator.

Alternatives	NPV	Adjusted NPV
Alternative A system	19,421.49	35,472.30
Alternative B system	29,246.64	29,246.64

Instead of extending the lives of the investments to a common denominator, another technique involves assuming that the alternative investments may be replicated forever. Alternative investments that can be replicated forever have infinite lives; thus the alternative investments have equal lives (or infinite lives). The NPV of an alternative that can be a *NPV replicated forever* (NPV_{rf}) equals the NPV of the *n*-year life project multiplied by an adjusted annuity factor and can be expressed as:

$$NPV_{rf} = NPV(n) \left[\frac{(1+r)^n}{(1+r)^n - 1} \right]$$

where *NPV(n)* is the NPV of the *n*-year investment, *n* is the length of the investment and *r* is the discount rate (for derivation see Copeland and Westland, 1988). Given the above problem with two alternative computer systems of unequal lives presented in Table 9, the NPV replicated forever for each alternative is shown in Table 11.

Table 11. NPV replicated forever.

Alternatives	NPV	NPV_{rf}
Alternative A system	19,421.49	111,904.77
Alternative B system	29,246.64	92,264.61

Based on these computations, Alternative A System should be selected over Alternative B System. When it cannot be assumed that alternative investments are replicable, the unadjusted NPV should be used to compare alternatives.

As a side note, the present value analysis has not considered inflation thus far. In most situations it can be assumed that inflation is considered already in the opportunity cost of capital and decision makers may adjust the cash flows for inflation. For further consideration of inflation see Copeland and Weston (1988) and Brealey and Myers (2000).

What is Return on Investment Methodology?

Return on investment (ROI) methodology is another technique traditionally used in capital budgeting decisions where the rate of return of an investment is compared to the opportunity cost of capital. In the previously mentioned survey by Deloitte & Touche, out of the 200 Chief Information Officers who were asked on how they measure the value of their IT investments, 43 percent said ROI was used (IT Value, 2003).

Other research shows that investments in e-commerce is also driven by ROI (Grant, 2002).

The return on an investment is calculated as the profit of the investment divided by the cost of the investment. If the return from the investment is greater than the opportunity cost of capital then the investment is worth more than it costs and should be undertaken. The *opportunity cost of capital* may be thought of as the expected return forgone by investing in the technology rather than in an equally risky investment in the capital market.

Let's evaluate a technology investment that costs $100,000 and will return $115,000 at the end of one year. Let's assume that this investment has similar risk to that of a security in the capital market with a return of 12 percent. Return is calculated as follows:

$$\text{Return} = \frac{\text{profit}}{\text{investment cost}} = \frac{115,000 - 100,000}{100,000} = .15 \ or \ 15\%$$

The return of the investment is 15 percent, which is greater than the opportunity cost of capital of 12 percent and thus the investment in the computer technology should be undertaken. For independent investment, the rules for the ROI methodology are as follows:

1. If return is greater the opportunity cost of capital, then make the investment.
2. If return is less than or equal to the opportunity cost of capital, then do not make the investment.

The problem with this methodology is that this return is only the true return if the cash flows are realized in two periods or less. When there are more than two periods it is questionable whether this method of calculating return yields the true return. In situations with more than two time periods some researchers suggest using the "internal rate of return." Due to this problem, ROI methods tend to be used as a supplement to other methodologies (Northrop, 2003).

ROI strategy for enterprise resource planning systems

Many firms make major IT investments in the form of *enterprise resource planning* (ERP) systems. ERP systems link all areas of a business into one management information system (Chase *et al.* 2004, pp. 452-462). This type of investment decision is highly risky and usually involves the decision and adoption of one major system, with many supplemental subsystem IT decisions. Great care must be exercised in developing a strategy for implementation, not just the ERP but how its ROI should be incorporated into the system and measured so an accurate picture of the IT investment and what it brings to the organization can be ascertained.

A strategy for capturing ROI within the context of an ERP can be a part of the implementation of an ERP (Boyle, 2003; Rao, 2000; Schniederjans and Cao, 2002, pp. 108-109). These steps basically include the following guidelines:

1. *Align your ROI to measure the criteria that is related to the implementation.* That is, if you are reengineering software, make those changes a part of the ROI computations.
2. *Assess possibilities for improvement.* This can be done by performing engineering analysis of service metrics, information transaction processing capabilities, activities required for procedures, and the various IT systems. This can also be performed by comparing internal benchmarks against expected or best practices observed in the firm and observed external to the firm. It is important to establish a set of service metrics that can be used to measure and reflect where the firm is today, prior to the implementation of the ERP or IT investment.
3. *Identify the opportunities for change within the context of the service and cost drivers the firm faces.* It is important to categorize them by changes to process, human resources, and technology.
4. *Capture benefits as early as possible during the implementation process.* This includes enlisting the aid of outsourcers, vendors and suppliers who might see changes more easily from the outside.

5. *Pay particular attention to software modifications.* Be sure to include all costs, including the technology, human resource, and other system change costs that may have to be included. Cost benefit analysis is suggested here as an ideal methodology.
6. *Sequence the implementation activities to better capture the benefits.* Sometimes new or unexpected benefits can accrue and should be included in the final ROI analysis.
7. *Build the capturing of benefits into the system by making it a routine part of the status reports.* Follow-through is critical in making the ROI data useful in the final analysis. Make sure to explore and measure all possible impacts of the new system on cost structures.

In summary, narrow the focus of the measures for ROI to those critical success factors that are related to firm success. Once these measures are in place, make sure to sequence implementation activities and aggressively collect the measure to insure comprehensive treatment of the benefits of the ERP system.

What is Internal Rate of Return?

Internal rate of return (IRR), also called *discounted-cash-flow rate of return*, is the discount rate that makes the NPV of a project equal zero. IRR is an extension or special case of the net present value methodology where the IRR is the rate that equates the present value of the cash flows with the initial investment. IRR may be used to evaluate independent or mutually exclusive investments. For an independent investment, if the IRR is greater than the opportunity cost of capital then accept the project; if not reject. When selecting one alternative investment among a mutually exclusive set, the investment with the highest IRR is selected; however, as expressed below, some caution is advised in using IRR to evaluate a mutually exclusive set. IRR may be calculated as follows:

$$NPV = C_o + \frac{C_1}{1+IRR} + \frac{C_2}{(1+IRR)^2} + ... + \frac{C^n}{(1+IRR)^n} = 0$$

Calculating IRR manually is a trial-and-error process. Potential IRRs are plugged into the above equation and adjusted as necessary to achieve a NPV of zero. The trial-and-error process may be unnecessary as most financial calculators are programmed to perform the required calculations. Table 12 provides an example using the formula above for evaluating an independent investment.

Table 12. Data for an internal rate of return problem.

	Co	C1	IRR %	NPV at 10%
Computer system A	-20,000	40,000	100	16,363.64

For an independent investment, the following rules may be used to guide the investment decision:

1. If IRR is greater than the opportunity cost of capital, then make the investment.
2. If IRR is less than or equal to the opportunity cost of capital, then do not make the investment.

According to these rules the investment in Computer System A in Table 12 should be undertaken if the IRR of 100 percent is greater than the opportunity cost of capital of 10 percent. Notice that these IRR rules and the NPV rules afford the same solution. The IRR rules will give the same answer as the NPV rules if NPV is a smoothly declining function of the discount rate for evaluating an independent investment. In other words, the NPV must decrease as the discount rate increases. If this does not hold true then the IRR should not be used and NPV provides the correct evaluation.

IRR will also give the same answer as NPV for a evaluating a mutually exclusive set if the alternatives in the set have the same initial investment size and the same lives. If one of these is different for the alternatives then IRR and NPV may give conflicting results. To illustrate, suppose there are two alternative computer systems in a mutually exclusive set and the initial investments are unequal for the two alternatives as presented in Table 13.

Table 13. Data for conflicting internal rate of return and net present value solutions.

Alternatives	C_0	C_1	IRR %	NPV at 10%
Computer system A	-20,000	40,000	100	16,363.64
Computer system B	-40,000	70,000	75	23,636.36

Table 13 also shows the IRR and NPV of the two alternative computer systems. According to the IRR rules, Computer System A is the best choice because it has the highest IRR, however, Computer System A also has the lowest NPV. IRR suggests Computer System A while NPV suggests computer System B; thus the two methodologies offer conflicting results. In situations where the initial investment is of different size the NPV methodology should be used in conjunction with the PI. Conflicting results may also arise in situations where the alternative investments have different lives. Again it is recommended that NPV be adjusted for unequal lives and used instead of IRR.

Several other problems exist with IRR. First, there may be more than one IRR that equates the NPV of investment to zero. According to Descrarte's "rule of signs", there may be as many different solutions to a polynomial as there are changes of sign. Thus, in situations where the sign of the cash flows changes there may be as many different solutions as there are sign changes. There are also situations where no IRR equates the NPV of an investment to zero and therefore IRR does not exist. In addition, there are cases when there is only one sign change but two IRRs that equate NPV to zero. Under these circumstances, the NPV method is much more reliable in offering the best solution and it is recommended for use.

Second, in many situations the opportunity cost of capital may not be equal for each cash flow. Thus far we have simply assumed the opportunity cost of capital is equal for all cash flows; however, in most situations this is not the case. When the opportunity cost of capital is not equal for all cash flows, the question of which opportunity cost of capital should be utilized to evaluate the alternatives is another problem of the IRR methodology. Some suggest using a complex weighted average of all the opportunity costs of capital to compute one IRR; however, we suggest using the NPV method. Finally, as mentioned before, NPV must be a declining function of the discount rate for it to provide valid results. In a situation where this does not hold true, NPV is a more applicable methodology.

Despite these problems, it seems that IRR is a widely used method because supposedly, management easily understands the percentage. However, it is essential to be fully aware of the problems associated with IRR and to realize when it can be used effectively. Consequently, It is recommended that IRR be used in conjunction with NPV. Ensuring that the two methodologies come to the same conclusion is a way to make certain the possible problems associated with IRR do not mislead the decision-making process.

Some of the advantages of using the IRR methodology that have appeared in the literature include the fact that it considers the time value of money, all cash flows, and yields a percentage management can understand. Some of the disadvantages include the need to estimate opportunity cost of capital, may yield multiple rates of return, may yield misleading conclusions for mutually exclusive projects, assumes cash flows may be reinvested at a return equal to IRR, and requires expertise for proper usage.

Computer-based solutions

A computer-based solution for IRR problems is available by using Microsoft© Excel©. By accessing Excel© and clicking the "Paste Function" key, a listing of "Function Categories" is provided. Clicking the "Financial" category reveals a "Function Name" listing, one of which is "IRR" or internal rate of return. Clicking on this function reveals a window with open boxes that allows an initial capital cost payment (entered as a negative value) and as many payments as is desired in the problem. By using the IRR Excel© program, the IRR can be computed easily. For example, if we invest say $1,000 and expect to receive $1,000 per year for the next three years, the entry data would be −1000, 1000, 1000, and 1000, resulting in a computed 84 percent as an IRR. The IRR program can also be used to simulate guesses on IRRs. This same Excel© program allows the decision maker an opportunity to include a guessed IRR and it seeks to simulate various percentages until the result is highly accurate.

A Word About Cash Flows

Each of the financial methodologies presented in this chapter is based on the assumption that cash flows and costs can be determined. Cash flows are most often based on quantifiable benefits that will arise from the investment in IT technology. However, many contend that the intangible benefits and costs of IT investments should also be considered in the evaluation and selection process. Many methods exist that consider these intangibles and are presented in the categorization of methods in this chapter and throughout this textbook.

Summary

The study of IT investment methodology requires understanding of many investment methodologies. In this chapter a brief description and/or listing of several dozen of the more commonly used IT investment methodologies were presented. While many of these methodologies will be discussed in greater length in later chapters, several of the most basic of the financial methodologies were presented in this chapter. These IT investment methods included present value analysis, return on investment and internal rate of return. The data requirements, computational procedures, and computer support functions were illustrated with example problems, along with problem exceptions, like unequal investment sizes and lives. In addition a variety of decision guild lines were presented to allow decision makers to customize their application and avoid common pitfalls in their use.

As was suggested in this chapter, there is a need to go beyond dollar values in truly and fairly evaluating an IT investment. Indeed, to really be inclusive, many investment planners insist that multiple criteria must be included in the analysis that combines both intangible and tangible criteria. In the next chapter we examine a classic financial methodology that has the ability to combine both intangible and tangible criteria called, "cost/benefit analysis."

Review Terms

Accounting rate of return
Balanced scorecard
Breakeven analysis
Cost benefit analysis
Cost displacement/avoidance
Cost effectiveness analysis
Cost revenue analysis
Discounted-cash-flow rate of return
Discount rate
Ex ante
Hedonic wage
Intangible criteria
Information economics
Internal rate of return (IRR)
Management science (MS)
Net present value (NPV)
NPV replicated forever (NPV$_{rf}$)

Objective criteria
Operations research (OR)
Opportunity cost of capital
Payback period
Portfolio management
Present value (PV)
Present value analysis (PVA)
PVA methodology
Profitability index (PI)
Post ante
Real options
Return on investment (ROI)
Return on management
Structural models
Subjective criteria
Tangible criteria
Time savings times salary

Discussion Questions

1. Why do you think there are so many IT investment methodologies?
2. Why were the methodologies categorized into the four different categories in this chapter?
3. Why do you think PVA is so often used with other IT investment decision-making methodologies?
4. What does the "opportunity cost of capital" really mean?

5. How can an "unequal investment life" impact an IT investment decision?
6. How can an "unequal investment size" impact an IT investment decision?

Concept Questions

1. What is the difference between "ex post" and ex ante" evaluations?
2. What is the difference between PV and NPV? Given an example.
3. What index can we use to compensate for unequal investment sizes?
4. What is ROI? How is it computed?
5. What is a reason why ROI is used as a supplement to other methodologies? Explain.
6. What is IRR? How is it computed?

Problems

1. Company XYZ wants to purchase a new IT, which will cost $120,000. The company will lease the equipment to a customer who has agreed to pay a leasing fee at the end of each of the next four years of $50,000. If the discount rate is 20 percent, is this a good investment for the Company XYZ? Use NPV analysis to prove your point.
2. Assume in Problem 1, the discount rate changes to 30 percent. Is this a good investment for Company XYZ? Use NPV analysis to prove your point.
3. An MIS manager has to chose one of two IT projects in which to invest (i.e., mutually exclusive investments). The initial cost for the two projects and their estimated income flows are presented in the table below. If the opportunity cost of capital is 15 percent and using NPV analysis, should they invest in either project? If they have to choose one, which of the two projects should they select?

	Project A	Project B
Initial cost	$500,000	$700,000
Income flow Year 1	$210,000	$125,000
Income flow Year 2	$240,000	$230,000
Income flow Year 3	$340,000	$655,000

4. (Refer to Problem 3) Assume now a 20 percent opportunity cost of capital. Should they invest in either project? If they have to choose one, which of the two projects should they select?

5. Assume you have an IT you want to invest in that costs $250,000 and will return at the end of the first year $265,000 in sales. What is its ROI?

6. The ABC Company bases its IT decisions on ROI. They have two mutually excusive computer systems (i.e., System A and System B) from which one must be chosen. Assume System A costs $1.5 million and System B costs $1.3 million. System A permits a contract worth $2 million the first year of its use. System B permits two contracts worth $750,000 and $1 million the first year of their use. Which system should ABC Company purchase? Use ROI to justify your answer.

7. A company has just computed their NPV on a cable system to support their information system architecture. The NPV is $2.5 million. If the investment cost for this system is $4 million, what is its profitability index?

8. A company must make a mutually exclusive choice between two different IT investment alternatives (i.e., A or B). Assume a discount rate of 15 percent. The initial cost and cash flow from the two projects are given in the table below:
    a. What are the NPVs for Alternatives A and B?
    b. Which alternative should be selected based on NPV?
    c. What are the PIs for Alternatives A and B?
    d. Which alternative should be selected based on PI alone?

	Alternative A	Alternative B
Initial cost	$60,000	$45,000
Cash flow Year 1	$10,000	$70,000
Cash flow Year 2	$40,000	$20,000
Cash flow Year 3	$90,000	$5,000

9. The company in Problem 8 now must make a mutually exclusive choice between the same two IT investment alternatives (i.e., A or B) with a discount rate of 30 percent. Given the same initial cost and cash flows on the two projects given above, answer the following questions:
   a. What are the NPVs for Alternatives A and B?
   b. Which alternative should be selected based on NPV?
   c. What are the PIs for Alternatives A and B?
   d. Which alternative should be selected based on PI alone?

10. A cutting-edge IT company is facing a classic decision as to which of three mutually exclusive choices they should select in the development of the next generation of programming software programs. They have three choices: to allow their internal IT staff to do the program, use some internal staff and some outsourced staff, or to completely outsource the program they are planning. If they have a discount rate of 10 percent and the initial cost and cash flows given in the table below, which should they choose? Use whatever analysis you want to defend your IT investment decision.

	Internal IT program	Mixed internal & outsource program	Outsource program
Initial cost	$130,000	$150,000	$220,000
Cash flow Year 1	$10,000	$30,000	$70,000
Cash flow Year 2	$50,000	$60,000	$70,000
Cash flow Year 3	$60,000	$70,000	$70,000
Cash flow Year 4	$60,000	$50,000	$70,000

References

Boyle, R.D., "Unlocking ROI," *APICS-The Performance Advantage*, Vol. 13, No. 6, 2003, pp. 36-39.

Brealey, Richard A. and Myers, Stewart C., *Principles of Corporate Finance*, 6th ed., New York, NY: McGraw-Hill, 2000.

Chan, Y.E., "IT Value: The Great Divide Between Qualitative and Quantitative and Individual and Organizational Measures," *Journal of Management Information Systems*, Vol. 16, No. 4, 2000, pp. 225-261.

Chase, R.B., Jacobs, F.R. and Aquilano, N.J., *Operations Management for Competitive Advantage*, 10th ed., Boston, MA: McGraw-Hill/Irwin, 2004.

Copeland, T.E. and Westland, J.F., *Financial Theory and Corporate Policy*, Reading, MA: Addison-Wesley, 1988.

Farbey, B., Land, F. and Targett, D., *How to Assess Your IT investment. A Study of Methods and Practice*, Oxford, UK: Butterworth-Heinemann, 1994.

Grant, E.X., "How Much Is Too Much To Spend on E-commerce?" *E-Commerce Times*, April 29, 2002.

IT Value. *The Controller's Report*, Vol. 2, Issue 7, 2003, p. 7.

Northrop, R., "The Hidden Cost of ROI," *Intelligent Enterprise*, Vol. 6, No. 10, 2003, pp. 46-48.

Parker, M.M., Benson, R.J. and Trainor, H.E., *Information Strategy and Economics*, Princeton: NJ, Prentice-Hall, 1989.

Powell, P., "Information Technology Evaluation, Is It Different?" *Journal of the Operational Research Society*, Vol. 1, 1992, pp. 29-42.

Renkema, T.J.W and Berghout, E.W., "Methodologies for Information System Evaluation at the Proposal Stage: A Comparative Review," *Information and Software Technology*, Vol. 39, 1997, pp. 1-13.

Rao, S.S., "Enterprise Resource Planning: Business Needs and Technologies," *Industrial Management & Data Systems*, Vol. 100, No. 2, 2000, pp. 81-88.

Schniederjans, M.J. and Cao, Q., *E-Commerce Operations Management*, Singapore: World Scientific, 2002.

Chapter 6

Cost/Benefit Analysis

Learning Objectives

After completing this chapter, you should be able to:

- Describe the stages of a "cost/benefit analysis."
- Explain how "cost/benefit analysis" can be used in IT investment decision-making.
- List and identify tangible and intangible costs and benefits used in "cost/benefit analysis."
- Be able to use various financial investment methodologies within the context of the "cost/benefit analysis."
- Explain how the "payback period" method works to aid in the "cost/benefit analysis." Define what "sensitivity analysis" is as it is related to "cost/benefit analysis."

Introduction

In this chapter we introduce a classic financial IT methodology called "cost/benefit analysis." It is a methodology that seeks to overcome some of the limitations of the return on investment (ROI) methodology discussed in Chapter 5. It does this in part by considering the usual cost information found in ROI analysis, but combined with "benefits" (not considered in ROI). As such it considers the all important and often intangible "value-added" contributions of personnel and managers. By bringing into the analysis more relevant criteria, cost/benefit analysis has become one of the most commonly used and appreciated financial methodologies in IT investment decision-making.

What is Cost/Benefit Analysis?

Cost/benefit analysis involves the estimation and evaluation of the net benefits associated with alternative courses of action. This technique most often entails comparing the present value of benefits associated with an investment to the present value of the costs of the same investment. Cost/benefit analysis is a widely used decision-making tool in both public and private settings and for a wide variety of different problems, including IT investment decision-making (Brown, 2001; Farbey, *et al.*, 1992; Farbey, *et al.*, 1993; Ryan, 2002; Sassone, 1988).

Cost/benefit analysis involves identifying costs and benefits for each alternative investment, discounting the costs and benefits back to the present, and selecting the best alternative according to a pre-specified criterion. Cost/benefit analysis may be used to evaluate an independent investment and to select one or several among a set of independent or dependent investments. Cost/benefit analysis may be used for *ex ante* (i.e., before project analysis), *ex post* (i.e., after project analysis) and in *medias res* (i.e., in progress analysis) investment evaluations. Like most analyses, cost/benefit analysis involves a series of steps or stages. Figure 1 shows five common stages in conducting a cost/benefit analysis. These sequential stages include: defining the problem, identifying costs and benefits, choosing a criterion, comparing alternatives, and performing sensitivity analysis. Each phase is discussed in more detail in the following sections.

Define problem

Defining the problem is extremely important in any type of decision-making, including IT investment decision-making. Analyzing the problem and defining is the only way to allow for the appropriate alternative solutions to be generated. Problem definition involves an in depth analysis of the situation; investigating the needs and requirements of an IT. After analysis, the problem may be defined and alternative solutions may be identified. A well-defined problem includes a specification of the objectives for an IT investment and a plan to attain those objectives. Possible objectives for an IT investment may be improved customer service, enhanced inventory control, or better information. A well-defined problem also includes a plan to attain the objectives.

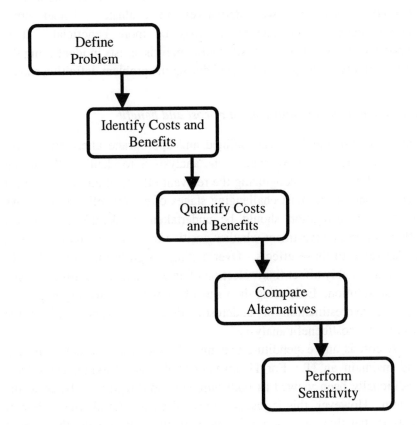

Figure 1. The five stages of cost/benefit analysis.

This part of problem definition involves generating all possible alternative courses of action and then if necessary narrowing this list down by eliminating unacceptable alternatives. Unacceptable alternatives may be ones that do not meet some basic constraint like those of a budgetary, legal, social, political, and/or institutional nature.

Due to the comprehensiveness of cost/benefit analysis, it tends to be a relatively expensive decision-making tool compared to other methodologies. By narrowing down the number of alternatives before conducting the analysis it is possible to better manage costs. Depending on budget constraints, it may be ideal to have a small number of alternatives, such as two alternatives, to evaluate. Under other circumstances, several alternatives may be more beneficial. It is important to remember that cost/benefit analysis assists in identifying the best alternative among a set selected during the problem definition stage.

Identification and quantification of cost and benefit

Once the problem has been defined and appropriate alternatives have been identified, the next stage in the analysis is to identify all relevant costs and benefits. Recognizing the relevant effects of an IT investment may be one of the most challenging stages of cost/benefit analysis. An intensive investigation should be undertaken to identify all relevant effects of an IT investment whether positive or negative, and to assign a dollar value to those effects. Overlooking a significant cost or benefit may unfavorably affect final selection of an alternative, resulting in a sub optimal solution. Estimating the value of an effect is also very important because overestimation or underestimation can adversely affect the results of a cost/benefit analysis.

A cost is any expenditure that must be incurred to procure, install, and maintain an IT. For IT investment decision-making, costs have traditionally been viewed as both *tangible* and directly attributed to the system. However, this is just one view of costs. An alternate view of costs is that they are *intangible*, meaning these are effects that cannot readily be assigned a value of the common unit of measure (usually dollars), and not directly attributed to an IT. Costs may be both tangible and intangible and both types should be evaluated in an IT investment decision where applicable. Current IT managers and controllers investment preferences strongly favor cost in any analysis for IT investment decision-making. According to a recent survey by Deloitte & Touche, out of 200 Chief Information Officers on how they measure the value of their IT investments, 81 percent said decreased costs were used (IT Value, 2003).

Table 1 shows examples of potential costs, tangible and intangible, that may be associated with a particular IT investment (Laudon and Laudon, 2004, p. 417; Farbey, *et al.*, 1992). It should be noted that the list is not intended to be a comprehensive, but provide examples of possible costs of typical IT investment decision-making situations. Each IT investment is unique in itself and with respect to the costs that should be included in the cost/benefit analysis.

Table 1. Potential costs of an IT investment.

Tangible:	Intangible:
Hardware	Resistance to change (change management)
Software	Inability to change
Telecommunications	Organizational restructuring
Needs specification and updates	Integration of new system into current situation
Services, e.g., installation, programming, etc...	Temporary loss of productivity (learning curve)
Personnel, e.g. hiring, training, etc...	Formulation of IT policy and controls
Running cost	Disruption to normal work practices
Furniture	Downtime

Identifying and managing intangible costs has not been given much attention in IT investment evaluation literature. However, intangible costs can impact the success or failure of an IT project and thus should be considered in cost/benefit analysis. Intangible costs, such as "resistance to change" or "inability to change", are not always considered in IT investment evaluations, when in fact they can have a major impact that may lead to the failure of a particular IT project. Incorporating factors like these when applicable into a cost/benefit analysis may not be a simple task but it is a necessary one. (Further discussion of incorporating intangibles into cost/benefit analysis may be found at the end of this section.)

A *benefit* is a positive consequence of undertaking an IT investment. Benefits often arise from making an improvement in the way an organization performs necessary tasks. Benefits, generally, may be classified into five categories:

1. Cost savings or avoidance;
2. Error reduction;
3. Improved operational performance;
4. Increased flexibility; and
5. Improved planning and control.

Table 2 shows examples of possible benefits of an IT investment decision situation (Laudon and Laudon, 2004, p. 417; Farbey, *et al.*, 1992). Benefits, just as costs, have been presented as tangible and intangible. Note that many of the common benefits associated with IT investments are intangible, meaning these are effects that cannot readily be assigned a value of the common unit of measure (usually dollars). Suppose that one of the major benefits of an IT investment is to improve customer satisfaction. Assigning a value to "improved customer satisfaction" may be a very difficult, if not impossible task.

There are several possible ways to manage intangible costs and benefits. One is to simply ignore them. In some situations, it may be acceptable and appropriate to leave intangibles out of the analysis because of the difficulty in assigning a value to them. It may also be determined that the intangibles do not have much of an affect and can easily be left out of the analysis. Another way to manage intangibles is to conduct the cost/benefits analysis without them but list them and describe their potential effects in an addendum. Here, intangibles are not directly included in the analysis but are presented as additional information to be considered when selecting the best alternative. A third way to manage intangibles is to utilize a surrogate measure for the intangible and include the effect directly into the cost/benefit analysis. A surrogate measure may be the value of a similar benefit or cost that is more easily assigned a value. Great care must be taken in selecting an appropriate surrogate measure to ensure that it provides a good approximation of the value of the actual benefit or cost. An example might be where quality costs in a company may have traditionally be one tenth the costs of equipment. Using the equipment costs to estimate or be a surrogate measure in this way provides a rough approximation, and where the same proportion is used in all alternatives being explored, might be a consistent means of including intangible quality costs. A fourth way to value an intangible is to conduct a survey to determine its value. Survey methods have been used extensively in cost/benefit analysis to determine the value of a cost or benefit. As an example, a survey may be designed to measure how valuable more timely information of an IT investment is to users. Users of the IT will be asked to assign a monetary value to the benefit and this value will be used in the cost/benefit analysis.

Table 2. Potential benefits of an IT investment.

Tangible:	Intangible:
Increased productivity	Improved asset utilization
Decreased operational costs	Improved resource control
Reduced workforce	Improved organizational planning
Lower computer costs	Improved organizational flexibility
Lower outside vendor costs	More timely information
Lower clerical and professional costs	Higher quality information
Lower in-house development costs	Enhanced organizational learning
Reduced rate of growth in expenses	Enhanced employee goodwill
Lower facility costs	Increased job satisfaction
Reduced software expenses	Improved decision-making
	Faster decision-making
	Lower error rates
	Improved operations
	Better corporate image
	Improved customer satisfaction
	Increased customer loyalty

One additional way of valuing an intangible is to use shadow prices. *Shadow prices* are measures widely used in economics to estimate the value of a good, or in the case of cost/benefit analysis, a cost or benefit. A shadow price is the value of an intangible, which indicates how much some specified index of performance could be increased (decreased) by the use (loss) of a marginal unit of that intangible. Shadow prices are used in situations when none of the aforementioned approaches will provide an appropriate value for an intangible. There is no comprehensive set of rules and procedures to determine shadow prices so it is necessary to employ an experienced analyst with good subjective judgment. There are several approaches to generating shadow prices. One is to use economic theory to determine shadow prices. A model is constructed based on common assumptions of economic theory and adjustments are made when there are violations to these assumptions to determine the value of an intangible. Another is to construct a model and use mathematical programming to generate shadow prices. Using this method, shadow prices are actually dual values generated by the algorithms in linear programming. (For further discussion on generating shadow prices using economic theory and linear programming see Sassone and Schaffer (1978).)

Compare alternatives

Once all costs and benefits have been identified and quantified into a common unit of measure, the alternatives are then compared to one another based on a common criterion. But before comparisons can be made, the costs and benefits that occur in subsequent time periods are often discounted back to today's dollars. In some instances, aggregate costs and benefits are compared without considering the time value of money; however, it is recommended that cash flows be discounted to account for this factor.

Discounting cash flows back that occur in subsequent periods is referred to as calculating the *present value* (PV) of a stream of cash flows. (Note the time value of money and present value was more fully discussed Chapter 5 of this textbook.) Calculating the present value is based on the basic principle of finance called the "time value of money". It is assumed that the value of monies or cash flows depends on the time period in which they are received. Cash flows received sometime in the future are worth less than those received today because those received today can be invested and begin accruing interest immediately. As a result, a discount rate must be selected and used to discount costs and benefits that occur in future time periods.

The present value of costs or benefits is calculated as follows:

$$PV = \sum_{t=0}^{n} \frac{A_t}{(1+r)^t}$$

where A_t is the cost or benefit in time period t, and r is the discount rate. The present value is the sum of the costs or benefits received in the future discounted back to today's value. The dis*count rate*, also called the *opportunity cost of capital*, is the rate that could be earned by investing in securities of comparable risk to that of the investment. An analyst or a member of the financial management team selects the appropriate discount rate based on the risk of the IT investment, and their expertise and knowledge of financial markets.

Suppose that an organization must select one computer system from a set of two alternative systems. A cost/benefit analysis is being used and decision makers have analyzed the problem, chosen alternatives and identified costs and benefits for each alternative. Tables 3 and 4 show the estimated costs and benefits of the two alternative computer systems. Assume the discount rate has been set at 8 percent for both computer systems.

The present value of the costs and benefits should be calculated for each alternative computer system. These values can then be used in calculations for selected criteria. The present values for the alternative computer systems in this example are presented in Tables 5 and 6. Notice that for both alternatives the present value of benefits exceeds the present value of costs, but the difference is larger for Computer System B.

Table 3. Costs and benefits of Computer System A.

	A_0	A_1	A_2	A_3
Costs:				
Hardware	10,000	1,000	1,000	1,000
Software	13,000	3,000	3,000	3,000
Services	2,000	1,000	1,000	1,000
Benefits:				
Increased productivity	--	10,000	6,000	6,000
Lower error rates	--	15,000	5,000	5,000

Table 4. Costs and benefits of Computer System B.

	A_0	A_1	A_2	A_3
Costs:				
Hardware	5,000	1,000	1,000	1,000
Software	10,000	5,000	0	5,000
Services	8,000	2,000	2,000	2,000
Benefits:				
Increased productivity	--	8,000	10,000	10,000
Reduced workforce	--	3,000	5,000	5,000

Table 5. Present value of costs and benefits for Computer System A.

Computer System A

Present value of costs	$PV = \dfrac{25,000}{(1+.08)^0} + \dfrac{5,000}{(1+.08)^1} + \dfrac{5,000}{(1+.08)^2} + \dfrac{5,000}{(1+.08)^3} = 37,885$
Present value of benefits	$PV = \dfrac{25,000}{(1+.08)^1} + \dfrac{11,000}{(1+.08)^2} + \dfrac{11,000}{(1+.08)^3} = 41,311$

Once the present values of costs and benefits have been calculated a criterion must be selected to determine which alternative is the best selection. Table 7 presents several criteria that may be used to select the best alternative. Let's look at each of these in light of this example.

Table 6. Present value of costs and benefits for Computer System B.

	Computer System B
Present value of costs	$PV = \dfrac{23,000}{(1+.08)^0} + \dfrac{8,000}{(1+.08)^1} + \dfrac{3,000}{(1+.08)^2} + \dfrac{8,000}{(1+.08)^3} = 39,330$
Present value of benefits	$PV = \dfrac{23,000}{(1+.08)^1} + \dfrac{15,000}{(1+.08)^2} + \dfrac{15,000}{(1+.08)^3} = 46,064$

Table 7. Common criteria to evaluate IT investments in cost/benefit analysis.

1. Maximize the ratio of benefits over costs
2. Maximize net present value of net benefits
3. Maximize internal rate of return
4. Shortest payback period

The first criterion that may be used is to select the alternative with the maximum ratio of benefits to costs. The *benefit/cost ratio* is the present value of benefits divided by the present value of costs and is calculated as follows:

$$\text{Benefit / Cost Ratio} = \dfrac{\displaystyle\sum_{t=0}^{n} \dfrac{B_t}{(1+r)^t}}{\displaystyle\sum_{t=0}^{n} \dfrac{C_t}{(1+r)^t}}.$$

The benefit/cost ratios for both alternatives in this example are presented in Table 8. The present value of costs and benefits for each alternative was calculated in Table 5, and according to these calculations, Computer

System B is the best alternative. The benefit/cost ratio for Computer System B implies that benefits of this alternative are 1.171 greater than its costs.

Table 8. Benefit/cost ratios.

	Computer System A	Computer System B
Benefit/Cost Ratios	$B/C = \dfrac{41,311}{37,885} = 1.090$	$B/C = \dfrac{46,064}{39,330} = 1.171$

The second criterion is to select the alternative having the largest *net present value of net benefits*. The net present value of net benefits is calculated as the present value of benefits minus the present value of costs discounted back to the present. The net present value of net benefits may be calculated as:

$$\text{Net Present Value} = \frac{B_0 - C_0}{(1+r)^0} + \frac{B_1 - C_1}{(1+r)^1} + ... + \frac{B_n - C_n}{(1+r)^n}$$

where B is the value of benefits, C is the value of costs, r is the discount rate, and n is the number of periods that benefits and costs occur. (See Chapter 5 of this textbook for a discussion of the use of computers to generate solutions and problems associated with NPV.) Table 9 shows that Computer System B is associated with a larger net present value of net benefits, $6,734, than that of Computer System A and is, thus, the better alternative. Recall from Table 6 that the present value of benefits and costs for Computer System B is $46,064 and $39,330. Using these present value calculations to determine the net present value of net benefits yields the same answer as those in Table 9 (46,064-39,330=6,734).

A variation of this criterion is to calculate the net present value without considering costs beyond the initial ones. In some situations it may be appropriate to assume that the costs beyond the initial costs are insignificant or nonexistent and should not be considered in the analysis. When there are no costs beyond the initial ones, NPV is calculated as the initial cost of the investment minus the present value of future cash inflows, or benefits.

Table 9. Net present value of net benefits for Computer Systems A and B.

Net Present Value of Net Benefits		
Computer System A	$\text{NPV} = \dfrac{0 - 25,000}{(1+.08)^0} + \dfrac{25,000 - 5,000}{(1+.08)^1} + \dfrac{11,000 - 5,000}{(1+.08)^2} + \dfrac{11,000 - 5,000}{(1+.08)^3} = 3,425$	
Computer System B	$\text{NPV} = \dfrac{0 - 23,000}{(1+.08)^0} + \dfrac{23,000 - 9,000}{(1+.08)^1} + \dfrac{15,000 - 3,000}{(1+.08)^2} + \dfrac{15,000 - 8,000}{(1+.08)^3} = 6,734$	

Internal rate of return (IRR) is a third criterion that can be used to evaluate an alternative(s) in a cost/benefit analysis. The internal rate of return is defined as the discount rate that equates the initial cost outlay with the present value of future cash flows. Alternatively, it may be defined as the discount rate that would make the NPV of an investment equal to zero. IRR is found by using trial and error to determine the rate that makes the NPV equal to zero. Financial calculators perform the calculations necessary to find the IRR. (See Chapter 5 of this textbook for a discussion of the use of computers to generate solutions and problems associated with IRR.) The IRR for Computer System A is 17.71 percent and for Computer System B the IRR is 25.87 percent. Final selection criteria might involve alternative investments with an IRR above a certain cutoff point or alternatives with the largest IRR maybe the best from a set of alternatives. IRR is considered to be an inferior criterion compared to the NPV criterion.

The fourth criterion is the *payback period*. Payback period is a common accounting and finance tool used to select the alternative that recovers its cost in the shortest amount of time. The payback period is the time when total investment is recaptured in cumulative cash flow. (For further discussion of the payback period method see Chapter 4 of the textbook.) Table 10 shows that the $25,000 initial cost of Computer System A is recovered in two years and the $23,000 initial cost of

Computer System B is also recovered in two years. According to the payback period criterion, both alternatives are equally good and either is acceptable. The major problem with the payback period criterion is that it does not consider the time value of money. In addition, the criterion may give illogical results when large cash flows occur in later time periods. The advantage of the payback period criterion is that it may be calculated quickly and requires no knowledge of present value calculations.

Table 10. Payback periods.

	Computer System A cash outflow and inflow	Cumulative cash flow	Computer System B cash outflow and inflow	Cumulative cash flow
Initial cost	25,000		23,000	
Year 1 cash flow	20,000	20,000	15,000	15,000
Year 2 cash flow	6,000	26,000	12,000	27,000
Year 3 cash flow	6,000	32,000	7,000	34,000

Two additional criterion exist that may be used in cost/benefit analysis. One is to maximize benefits for a given level of costs. That is, the alternative with the largest amount of benefits for a given level of costs is the best alternative. This rule is applicable when it can be assumed that each alternative IT investment has relatively the same level of costs. The other rule is to minimize cost for a given level of benefits. Here the benefits for each alternative must be relatively the same level, then the alternative with the lowest cost is the best alternative.

Sensitivity Analysis

Sensitivity analysis is defined as determining the reliability of the decision generated from a cost/benefit analysis. In cost/benefit analysis having the actual values of every cost and benefit associated with alternative investments would be ideal. If these values were known for certain there would be no error. However, the values of the costs and benefits, especially those intangibles ones, are only estimates of the true value and thus are associated with some amount of error. Performing a sensitivity analysis is one way to determine the degree of error in the estimates.

The degree of error in the estimates determines the reliability of the final NPV or value of whichever criterion is being employed, and thus the reliability of the decision yielded by the analysis. If the NPV criterion is being used, and the NPVs of the alternative investments are similar, like in the example above, and the degree of error is large, then it is difficult to be sure that the alternative with the largest NPV is actually the best. Alternatively, if the NPVs of the alternative investments are very different and the degree of error is small, then the decision yielded by the analysis is stronger and it is easier to select one investment over another.

There are many variations to performing sensitivity analysis, but a common way is to select costs, benefits, or other parameters in the NPV calculation with large amounts of error and vary them to examine their effects. The analysis may involve selecting high and low values of a parameter and assess the effects on NPV. The result is having a NPV associated with the original value, another NPV calculated with the high value, and another with the low value. The degree of dispersion of these NPVs shows how different values of a parameter affect the final NPV and corresponding decision. Varying just one parameter may change the highest NPV of one alternative to prefer a different alternative, making the results of the analysis unreliable.

One problem with the selective sensitivity analysis is that in an IT investment decision a large number parameters, including the discount rate used in the NPV calculation, may have a higher degree of error and thus be critical. If there are ten critical parameters with a high and low estimate of each, and two alternative investments, 40 additional NPV calculations must be computed and analyzed to determine their degree of sensitivity. This large amount of information can be condensed and displayed in an easy to read form by deriving a probability distribution of NPV outcomes reflecting all possible NPVs given variations in critical parameters. The result of this type of sensitivity analysis is a graphical depiction that reveals the chances of alternative investments breaking even, failing or succeeding. At a glance decision-makers are given a lot of information that can be easily processed and hopefully, better assist in the decision. In depth discussion of this type of sensitivity analysis is presented in Sassone and Schaffer (1978).

What is Cost/Effectiveness Analysis?

Cost/effectiveness analysis is another cost-analysis technique that considers costs and effects that are defined in different terms. Just as in cost/benefit analysis, the problem is defined, an objective is set, and alternatives are generated in cost/effectiveness analysis. The difference lies in the unit of measure. In cost/benefit analysis, alternatives are evaluated based on costs and benefits measured in monetary terms. In cost/effectiveness analysis, costs are evaluated based on monetary terms and benefits are gauged in terms of how effectively each alternative meets a common objective. Each alternative is evaluated based on its individual costs and its contribution to meeting the same effectiveness criterion. An example of an effectiveness criterion would be an objective like improving customer satisfaction or increasing organizational learning. The best alternative, the most cost effective one, would be the one that offers the lowest cost for any given increase in customer satisfaction or organizational learning. Cost/effective analysis may be an appropriate alternative methodology for the evaluation and selection of IT investments when intangibles are a critically important part of the analysis. For further discussion of cost/effectiveness see Levin and McEwan (2001) and for cost/effectiveness analysis in IT investment decision-making see Sassone (1988).

Summary

This chapter has presented the IT investment methodology of cost/benefit analysis. Cost/benefit identification and quantification methods were described. A variety of intangible and tangible costs and benefits where identified as a means of performing the analysis and as a means of recognizing the many opportunities this type of analysis has in IT investment decision-making. A discussion and illustration of comparison methods useful in conducting the cost/benefit analysis included present value analysis, net present value analysis, IRR, cost/benefit ratio, and other methods. The intent here was to show how the cost/benefit analysis was adaptable to a variety of other classically used financial evaluation methodologies for IT investment decision-making.

Cost/benefit analysis has been shown in this chapter to provide a good bridge between use of multiple criteria and combinations of intangible and tangible measures for IT investment decision-making. Not all problems can convert the multiple criteria used in its analysis into dollars, as cost/benefit analysis requires. The next three following chapters in Part III, "Multi-Criteria Information Technology Decision-Making Methods," will discuss a variety of methodologies used to incorporate multi-criteria that permit a wide range of measures for use in IT decision-making process.

Review Terms

Benefit/cost ratio	Medias res
Benefits	Net present value (NPV)
Cost/benefit analysis	NPV of the net benefits
Cost/effectiveness analysis	Payback period
Ex ante	Present value (PV)
Ex post	Sensitivity analysis
Intangible benefits	Shadow prices
Intangible costs	Tangible costs
Internal rate of return (IRR)	Tangible benefits

Discussion Questions

1. Is one stage of the "cost/benefit analysis" more important than another? Can we drop off any of the stages in a particular analysis?
2. Why is "cost/benefit analysis" more expensive than other analyzes?

3. Why would we use the "benefit/cost ratio" for a comparison in a "cost/benefit analysis"?
4. Why would we use the NPV for a comparison in a "cost/benefit analysis"?
5. Why would we use the IRR for a comparison in a "cost/benefit analysis"?
6. Why would we use the "payback period" for a comparison in a "cost/benefit analysis"?

Concept Questions

1. What are the five stages of the "cost/benefit analysis"?
2. "Cost/benefit analysis" can be used in what kinds of investment decisions?
3. How do you identify relevant costs in a "cost/benefit analysis"?
4. What are examples of costs in a "cost/benefit analysis"? Give examples of at least five.
5. How do you identify relevant benefits in a "cost/benefit analysis"?
6. What are examples of costs in a "cost/benefit analysis"? Give examples of at least five.
7. Once the costs and benefits are identified, how can we compare them in order to make a decision?
8. What determines the type of methodology used in the comparison during a "cost/benefit analysis"?
9. How does "sensitivity analysis" benefit "cost/benefit analysis"?
10. What is "cost/effectiveness analysis"?

Problems

1. You have four IT investment alternatives (i.e., A, B, C and D) in which one must be selected. The "benefit/cost ratios" for the four alternatives are 1.6, 3.4, 6, and 2.5, respectively. Based on these ratios, which alternative would you choose? Explain your answer.

2. A company must plan its electronic data interchange (EDI) equipment purchase around a combination of criteria, including, initial cost, yearly maintenance costs, labor savings due to increased productivity, and increased profits due to new services the new IT will provide. The company has collected the cost/savings/profit information and computed "benefit/cost ratios" for the three alternatives EDI systems (i.e., EDI(1), EDI(2), or EDI(3)). The resulting ratios are 1.1, 2.1, and 2.5, respectively. Based on the ratios, which of the three systems would you choose? Explain your answer.

3. One of two IT improvement projects must be undertaken to upgrade an existing intranet communications system. The estimated NPV of the benefits of System A is $450,000 over the life of the investment. The estimated NPV of the benefits of System B is only $300,000 over the same life of the investment as System A. The estimated NPV of the costs of System A is $270,000 over the life of the investment. The estimated NPV of the costs of System B is $105,000 over the life of the investment. Based on the benefit/costs ratio, which system should be selected? Show your work.

4. One of four IT software (i.e., Software's A, B, C and D) can be purchased to do the same job. The estimated NPV of the benefits of Software A is $100,000 over the life of the investment. The estimated NPV of the benefits of Software B is only $90,000 over the same life of the investment as Software A. The estimated NPV of the benefits of Software C is $80,000 over the life of the investment. The estimated NPV of the benefits of Software D is $75,000 over the life of the investment. The estimated NPV of the costs of Software A is $50,000 over the life of the investment. The estimated NPV of the costs of Software B is $48,000 over the life of the investment. The estimated NPV of the costs of Software C is $45,000 over the life of the investment. The estimated NPV of the costs of Software D is $40,000 over the life of the investment. Based on the benefit/costs ratio, which system should be selected? Show your work.

5. To compare two alternative IT investment alternatives, an IT manager has selected to use the "payback method". Assume the

IT will only last three years and be scraped at the end of the third year with the estimated proceeds of the IT sale at that time added to the cumulative cash flow. The table below shows the cumulative cash flows for the two alternative systems (i.e., A and B) from which one must be chosen. Which system should be selected? Explain your answer.

	Alternative System A	Cumulative cash flow	Alternative System B	Cumulative cash flow
Initial Cost	125,000		110,000	
Year 1 cash flow	70,000	70,000	50,000	50,000
Year 2 cash flow	40,000	110,000	50,000	100,000
Year 3 cash flow	10,000	120,000	20,000	120,000
Scrap value (end of life)	2,000	122,000	5,000	125,000

6. A company would like to choose the best two out of three possible IT investment alternatives (i.e., A, B, and C). To make the comparison, the "payback method" is chosen. Assume the IT will only last five years and be scraped at the end of the fifth year with the estimated proceeds of the IT sale at that time added to the cumulative cash flow. The table below shows the cumulative cash flows for the three alternatives from which two must be chosen. Which two systems should be selected? Explain your answer.

	Altern. A	Altern. B	Altern. C
Initial cost	300,000	350,000	370,000
Year 1 cash flow	120,000	150,000	180,000
Year 2 cash flow	100,000	100,000	160,000
Year 3 cash flow	78,000	50,000	60,000
Year 4 cash flow	50,000	20,000	40,000
Year 5 cash flow	0	10,000	30,000
Scrap value	12,000	12,000	10,000

References

Brown, M.M., "The Benefits and Costs of Information Technology Innovations: An Empirical Assessment of a Local Government Agency," *Public Performance & Management Review*, Vol. 24, No. 4, 2001, pp. 351-367.

Farbey, B. Land, F. and Targett, D., *How to Asses Your IT Investments. A Study of Methods and Practice*, Oxford: Butterworth-Heinemann, 1993.

Farbey, B., Land F. and Targett, D., "Evaluating Investments in IT," *Journal of Information Technology*, Vol. 7, 1992, pp. 109-122.

IT Value. *The Controller's Report*, Vol. 2, Issue 7, 2003, p. 7.

Laudon, K.P., and Laudon, J.P., *Management Information Systems: Managing the Digital Firm*, 8th ed., Upper Saddle River, NJ: Prentice Hall, 2004.

Levin, H.M. and McEwan, P.J., *Cost-Effectiveness Analysis*, 2nd ed., London: Sage Publications, 2001.

Ryan, S.D., "Information-technology Investment Decisions: When Do Costs and Benefits in the Social Subsystem Matter?," *Journal of Management Information Systems*, Vol. 19, No. 2, 2002, pp. 85-93.

Sassone, P.G. and Schaffer, W. A., *Cost-Benefit Analysis: A Handbook*, New York, NY: Academic Press, 1978.

Sassone, P.G., "A Survey of Cost-Benefit Methodologies for Information Systems," *Project Appraisal*, Vol. 2, 1988, pp.73-84.

Part III

Multi-Criteria Information Technology Decision-Making Methods

Chapter 7

Critical Success Factors, Delphi Method and the Balanced Scorecard Method

Learning Objectives

After completing this chapter, you should be able to:

- Define "critical success factors".
- Explain how critical success factors are used in organization strategic planning and IT investment decision-making.
- Explain how to develop a set of critical success factors.
- Explain what the "Delphi method" is and how it can be used in IT investment decision-making.
- Explain what the "balanced scorecard method" is and how it is related to strategic planning.
- Define the four perspectives of the "balanced scorecard method" and how they are related.
- Explain how the "balanced scorecard method" can be applied.

Introduction

In the previous three chapters of Part II, we focused on financial methods for *information technology* (IT) investments. This chapter is the first of three chapters in Part III, "Multi-Criteria Information Technology Decision-Making Methods" that redirect the focus of methodology to include broader issues of IT investment decision-making. That is, rather than focusing on one piece of IT or a single system decision or evaluation as chiefly a financial analysis, we look here at decision-making in general and investing in larger portions or whole systems. Many of the methodologies contained in Part III can also be used to make individual IT or subsystem decisions, but they have broader application and are generally more inclusive of decision-making criteria. This is why these methodologies are identified as "multi-criteria" in nature.

In this chapter we primarily introduce three conceptual methods called "critical success factors", the "Delphi method", and the "balanced scorecard method" as approaches to beginning a process of large-scale, corporation-wide planning and decision-making for IT. Unlike the quantitative methods in previous chapters, these are conceptual methods ideal for use in beginning the process of identifying information needs on which to base IT acquisition decisions.

What are Critical Success Factors?

In Chapter 1 we presented a multi-step *management information systems* (MIS) hierarchical planning approach to IT systems. In Figure 1, the strategic planning steps from the MIS hierarchical planning approach are again presented. One of the methodologies commonly used in these strategic planning steps is called "critical success factors" (Digman, 1990; pp. 247-253; Laudon and Laudon, 2004, pp. 380-382; Young and O'Byrne, 2001, pp. 269-303). *Critical success factors* (CSFs) are a set of requirements that if a firm achieves them, they are assured of business success. CSFs are sometimes referred to as *cost drivers* or *value drivers*, which are activities that have an impact in reducing costs or adding value to service products offered to customers (Young and O'Byrne, 2001, pp. 270-274). CSFs can include the activities of research and development, new product development, customer service, quality service, etc. They are commonly a part of all the steps in the MIS hierarchical planning approach in Figure 1.

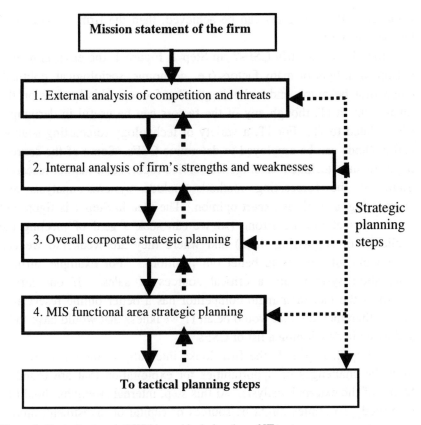

Figure 1. Strategic steps in MIS hierarchical planning of IT systems

As we can see in Figure 1, the firm usually starts with its *mission statement* (i.e., a document that states its general or broad objectives in conducting its business with the external environment of stakeholders (i.e., customers, stockholders, employees, suppliers, etc.). When used in the context of the MIS hierarchical planning of IT systems, CSFs are identified both externally to the organization and internally (i.e., Steps 1 and 2 in Figure 1). CSFs are examined at the level of the firm, the industry, and the general environment. Basically this approach argues that an organization's information requirements are determined by a number of CSFs that are viewed as goals to be achieved by MIS managers. If these goals can be achieved, then the firm is assured of business success.

How do we identify CSFs? In Step 1, Figure 1, the environment is examined in light of many factors (i.e., economic, sociological, political, competitor behavior, and technology). Our focus in this book is on the investment in IT, though any of the factors can be useful in deriving a CSF related to IT. For IT, a variety of technology forecasting sources and methods can be employed to determine CSFs. Some of the sources might include technical intelligence service reports and market research. Some of the forecasting methods include systems analysis and engineering, as well as expert opinion. The idea in Step 1 is figure out where competitors are strong (so we can seek a goal of meeting the challenge or surpassing them) and where they are weak (so we can exploit our advantages to better our position). For example, airlines reservation systems are a critical sources of sales. If one airline identifies the fact their main competitor has a better online reservation system then they do, that would be an IT weakness that would have to be addressed in developing a list of CSFs.

In Step 2, Figure 1, the firm looks internally at their resources to meet the challenges or opportunities for exploitation that are found in Step 1 of the external analysis. At this step, internal strengths should be identified. Examples might be sources of capital of investment, unique assets the firm owns, like patents, and outstanding human resources. While the airline firm mentioned in the paragraph above might not be able to make a quick investment in their online reservations, they might be able to bolster existing phone and airport reservation systems to meet their competitive challenger who has the better online system.

In both Steps 1 and 2, an organization learns what they do well and what the competition is doing well. Some of those actions might generate substantial profit or provide a competitive advantage in reducing costs. Those actions may directly impact critical factors in the success of the organization. Hence, they become CSFs.

In general there are at least five criteria as presented in Table 1 that can be used as a guide in the identification of the CSFs. While these guides are general enough to be used in all the functional areas within and for the organization as a whole, they can also be thought of as relating specifically to IT investment decision-making. To illustrate the "relationship with strategic goals" CSF criterion in Table 1, an example is presented in Table 2.

Table 1. Criteria useful in determining CSFs.

CSF criterion	Examples
1. Impact on performance measures	Impact on sales, profitability, cash flow, etc.
2. Relationship with strategic goals	Relationship with goals of differentiation, cost minimization, market segmentation, etc.
3. Relationship to life-cycle stage of products or business	Relationship in a state of introduction, growth, maturity or decline for products
4. Relationship with major business activities	Relationship with major customers, suppliers, etc.
5. Relative size of investment	How much money is needed now and in the future

Note in Table 2, a single overall strategic plan is often broken down into multiple functional MIS plans, goals and CSFs. Also note in Table 2, that at the overall strategic level "quality" could be viewed as a CSF for the organization as a whole, while the functional MIS has a related but more diverse set of CSFs (i.e., timely information, cutting-edge IT, and effective marketing intelligence). This makes the point that CSFs can vary at each level in the organization, making their identification difficult. On the other hand, CSFs identification is essential for many firms if they are to be competitive. As such many firms employ polling and analyzing their manager opinion on what constitutes a CSF as suggested in Figure 2, to identify the organization's CSFs. The outcome of this polling process can be clear direction on the future development of management information systems, like decision support and database systems that will directly support the desired CSFs. One of the polling methodologies commonly used when managers, executives, or experts are used to identify CSFs is called the "Delphi method". (We will discuss this methodology in the next section of this chapter.)

Table 2. Examples of CSFs relationship with strategic goals.

Overall corporation strategic plan	MIS strategic plans	MIS goals	Critical success factors
Develop the highest quality in services offered in industry	Develop an MIS that supports managers ability to monitor changes in service quality	Achieve employee systems to continually monitor and report service success and failures	Timely information (so problems can be identified quickly and corrections undertaken)
	Develop an MIS that supports development of new high quality service products	Achieve systems to engineer and analyze new product quality	Cutting-edge IT (to permit highest levels of accuracy in engineering and the finished product)
	Develop an MIS that supports the customer's idea of quality	Achieve customer survey systems to provide access to marketing personnel to make improvements	Effective marketing intelligence (from perhaps an online survey system so new suggestions can be effectively implemented)

In summary, CSFs are used for a number of areas of planning. As mentioned, they can be used to identify where the firm has strengths to exploit, weakness to build up, and eventually with polled executive opinion, a more detailed definition of specific MISs useful in supporting the CSFs. In this sense, CSF acts as a conceptual guide to finding opportunities to IT investment. As Wen *et al.* (1998) suggests CSFs can be a useful tool for IT investment decisions where general guidance on technology is preferred over the worst excesses for conceiving accurate but meaningless numbers from the financial types of IT investment methodologies. (In Chapter 8, we will discuss the use of multi-factor scoring methods that can be used to quantitatively evaluate the importance of CSFs and developing their prioritization for purposes of their implementation.)

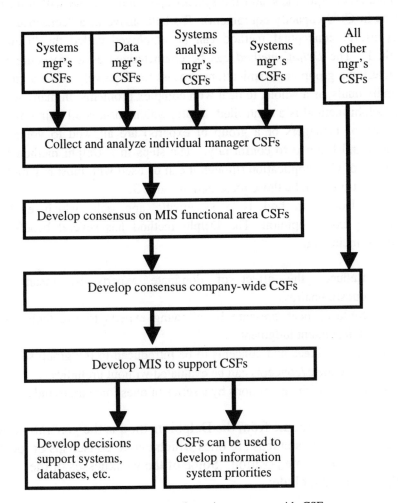

Figure 2. Polling company managers to determine company-wide CSFs.

What is the Delphi Method?

In Figure 2, it is necessary to "Develop a consensus company-wide CSFs". This can be a difficult task in IT investment decision-making since differing functional areas, and even differing departments in the MIS area, might have differing ideas on what CSFs they want, as it's related to their specific area (e.g., the data store staff will want to invest in data store equipment and the system operations people will want to invest in CPU capacity equipment). To help arrive at a "consensus" a very useful conceptual method can be employed called the "Delphi method". The *Delphi method* can be characterized as a procedure for structuring a group communications process to effectively allow a group of individuals, as a whole, to deal with complex problems. In many ways the Delphi method is a controlled debate, which ensures all opinions (of a group of managers or a group of experts) are allowed to voice an opinion and bring it to conformity. The steps in a Delphi method for purposes of CSF application (though it can be used with most any multi-criteria setting) can be those presented in Table 3.

The Delphi method is a systematic approach, which evokes collective expert opinion. The Delphi method has several beneficial features, including:

1. Reduces the affect of dominate experts by means of questionnaires.
2. Reduces peer pressure by allowing experts to use their own independent judgment .
3. Allows ideas and concepts to be introduced to the group so that these ideas/concepts can be evaluated without prejudice.
4. Reduces fringe opinions by a series of questionnaire rounds.

A key outcome from the Delphi method is the generation of ideas; whether these ideas are ones which evoke consensus or ideas which are at extreme position. Some of the weaknesses of the Delphi method include:

Table 3. Steps in the Delphi method for CSF consensus building.

Steps	Explanation/description
1. Select a group of managers (experts) to determine the CSFs	These experts should be knowledgeable or have expertise in all areas of the firm. They should also be kept anonymous from one another.
2. Send each expert in the group a questionnaire requesting what they feel are appropriate CSFs	The questionnaire should clearly state all necessary parameters that are required to complete it (i.e., time frame on which to return the questionnaire; the level in the firm –strategic, tactical, and operational; etc.).
3. Collect the questionnaires from the group document results, analyze, and prepare a report.	The idea here is to see what CSFs are in common with the experts. Create distributions to define the frequency of selection by expert for each CSF. Prepare a report that summarizes the selections and the frequency of selection by expert.
4. Send the report and a revised questionnaire to the experts.	Ask the experts to use the information in the report to update their selections of CSFs. Ask the experts whose suggested CSFs are in the minority to explain why they don't feel the need to change and conform to what the majority of experts feel about what the CSFs should be.
5. Repeat Steps 3 and 4 until changes in the CSFs stop.	The idea here is to permit the experts to see how their selections are being supported or not supported by the others in the group. Eventually, the majority will form some consensus and changes in the set of CSFs chosen will no longer take place. It is important to allow the minority experts to share reasons why they choose not to change their opinion since this can sometimes sway the group as a whole.

1. It is difficult to perform and select experts.
2. The questionnaires must be meticulously prepared and tested to avoid ambiguity.
3. The time requirements to do the rounds of polling the experts.
4. Is only as valid as the experts who make up the group.
5. Ignores disagreements so that an artificial consensus might be achieved.

6. It does not always produce more accurate answers than other methods.
7. It fails to meet the requirements required for scientific research and that the process is weakened by not allowing the experts to discuss issues.
8. It has validity and reliability issues.

Some of the strengths of the Delphi method include:

1. Enables issues to be explored in an objective fashion.
2. It may be most useful where opinions are being sought and where there is little or no role for evidence.
3. It may the best way to explore alternatives, and the pros and cons for each alternative.
4. It can utilize existing staff of a company, who may be more expert than outside consultants.

The Delphi method can be used for a wide-range of forecasting and decision-making applications (Laudon and Laudon, 2004, pp. 471-472). The Delphi method has been specifically suggested as an IT investment methodology by Wen *et al.* (1998).

What is the Balanced Scorecard Method?

The *balanced scorecard method* is a technique companies use to translate their strategies into objectives and measures (Kaplan and Norton, 1996a; Norton, 1996b). It moves away from evaluating performance based solely on financial measures and incorporates both financial and non-financial performance measures (e.g., quality and customer satisfaction when judging performance) (Kaplan and Norton, 1992; Kaplan and Norton, 1993; Matinsons *et al.*, 1999). It helps to translate these objectives to give management a complete picture of operations and to communicate company goals and strategies to all levels of the business.

All members of an organization are enabled to align their goals with the company's and with the common goals of the business known to all members; this enables them to work more efficiently and effectively together to achieve the goals. Most importantly, the scorecard allows companies to evaluate whether they are meeting their objectives, based on both financial and non-financial measures using tangible and intangible assets.

Sadly, it is true that financial measures may encourage managers to focus more on short-run decisions, than long-term decisions. Decisions that could improve the appearance of the manager's performance in the short-term, but not increase company value in the long-term. This is particularly true for IT decisions since they are longer-term capital investments. For example, when managers are evaluated on financial matters alone, they may make decisions that are not in the best economic interests of a company, such as holding on to an asset in the current period in order to sell it in a period when a boost in income is needed. The balanced scorecard's interrelated parts give managers the ability to perform to the best of their abilities and be rewarded on their successes in more areas other than return on investment aspects of financial measures. The balanced scorecard provides for performance measures rewarding managers for areas such as customer satisfaction. They also encourage managers to more comprehensively consider a multitude of criteria for business performance success.

Balanced scorecard components

The balanced scorecard is organized into four areas or *perspectives*: financial, customer, internal business, and learning and growth as presented in Figure 3 (Kaplan and Norton, 1996b; Young and O'Byrne, 2001, pp. 269-303; Zee, 2002, pp.170-210). Note that each of the four scorecards are interconnected and related to the company's strategic planning. Each of the four in Figure 3 represents a set of criteria or measures used to specifically evaluate the company's progress from where their actual performance currently places them in the accomplishment of the strategic goals and a clear target stated where they would like to be (presumably at a higher level of performance). The basic idea is to balance the multiple measures of business performance in each of these scorecards so all the stakeholders (i.e., stockholders, customers, employees, suppliers, etc) have an improved position over time and that position will reflect the organization's strategic planning goals. Let's describe at each of these four scorecards in greater detail.

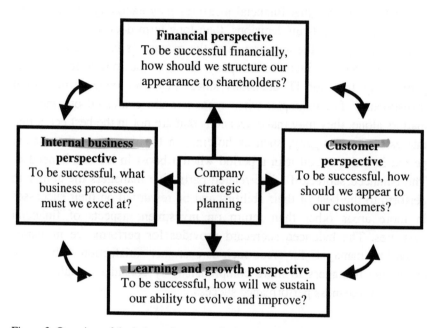

Figure 3. Overview of the balanced scorecard's four perspectives.

— *Financial Perspective*: The financial perspective evaluates how a company is meeting its objectives through financial measures. The financial measures focus on action that has already taken place. This perspective focuses on the shareholders and what steps the company is taking to ensure that they are meeting their shareholders' financial expectations. Typical financial measures that enable a company to focus on the bottom line are operating income, return on capital, and economic value added. It is important for all the measures to be tied to relevant strategic goals the firm seeks to achieve. Therefore, any type of financial measure (e.g., internal rates of return, return on investments, etc.) can be used. An example of the financial perspective scorecard in the area of developing IT projects is presented in Table 4.

Table 4. Example of financial perspective scorecard for developing IT projects.

Goals	Measure	Actual performance	Targeted performance
Maintain financial control	Costs of maintenance as a percentage of maintenance and project development costs	15%	12%
Being profitable in developing IT projects	Profit per employee	$2,700	$3,000
Being cost effective	Ratio of budgeted project costs to actual project costs	1.1	1.0

Note in this example that we are developing a balanced scorecard for just IT project development, which is at the MIS functional level of an organization. One of the major advantages of this methodology is that it can be used in all areas of a firm to break down the strategic goals into functional and departmental levels, thereby communicating important strategic goals to everyone in the organization. Indeed, this methodology can be used at the strategic, tactical and operational levels to plan IT change, develop IT implementation strategies, and set objective measured goals in which to monitor progress. By comparing the "actual performance" with the "targeted performance" managers can easily see their "score" or how well they are doing.

Now, there is a concern about the balanced scorecard method that should be brought up here. That is, if managers focus their effort exclusively on improving the financial scorecard to the exclusion of all other considerations, they can easily cause other areas of the business to not fair so well in their scoring. Fortunately, this is the beauty of the balanced scorecard method; we seek to achieve a "balance" between the scorecards in other areas so everybody ends up improving their respective operations. Such a balance is not just based on objective financial numbers but also on non-financial subjective criteria that can be measured and used in the scorecards. One of the most important areas for long-term success is considering the customer.

Customer Perspective: The customer perspective is the core perspective which defines how a company differentiates itself from the competition to attract, retain, and deepen relationships with targeted customers. The customer perspective identifies the customer segment that the industry will focus on. The business's performance is then based on how the company is exceeding these customers' expectations. The balanced scorecard will use measures customized to their specific target segment such as customer satisfaction, customer retention, new customer acquisition, customer profitability, and market share to evaluate its performance in the customer perspective. Within the customer perspective, a company should focus on product and service attributes, customer relationship, and image and reputation in order to evaluate whether or not they are meeting their strategy. Listing items such as customer satisfaction and customer service on a balanced scorecard can help alert managers and employees to their importance. They can then be guided to take action to enhance those features (e.g., providing quality products at a value price).

There are several different avenues a company can take in order to achieve its objectives in the customer perspective. According to Kaplan and Norton (1996a) companies should practice operational excellence, customer intimacy, or product leadership. When a company makes a decision about which position best fits them they should excel in that area while maintaining an acceptable level in the other two. If operational excellence is their objective then the company should excel at product pricing, while maintaining an acceptable level of customer service and product quality.

There are many possible measures that can be used to gauge the customer's perspective. In addition to those in Table 5, other measures might include: processing costs as a percentage of total business, data processing costs per worker, per job, per batch of jobs, etc. In the area of customer quality there is literally thousands of measures that can be included but the idea is always to find those that relate to the current strategic goals the firm wants to achieve.

Table 5. Example of customer perspective scorecard for developing IT projects.

Goals	Measure	Actual performance	Targeted performance
Being responsive	Response time at terminals for online transactions	0.15 minutes	0.10 minutes
Providing quality service	Availability of system to serve project needs	22.9 hours/day	24 hours/day
Providing quality service	Customer satisfaction score (1 to 10, where 10 is perfect)	score of 6	score of 8
Providing quality service	Number of project or business transaction failures per week	5/week	0/week
Being cost effective	Processing costs per online project or business transaction	$0.0002	$0.0001

Internal Business Perspective: The internal business perspective identifies what internal business processes the company must excel at in order to be successful. (This is the same as determining the CSFs but focuses specifically on production processing.) These are processes that have the greatest impact on delivering value to the customer and satisfying shareholder expectations of financial returns. They may be processes that the company is currently not performing, but through the balanced scorecard are identified as necessary in order to meet their competitive goals.

The internal business perspective identifies three areas companies can exceed in, which include innovation, operations and post sales services. Innovation identifies customer's needs and wants, and then develops products to meet those needs and wants. An operation is how the business produces and delivers their products and services to their customers. Post sales service is how the customers are serviced after purchase. These processes are measured using gauges for quality and delivery time.

Results from the internal business perspective may be obvious in some cases and vague in others. Improving process efficiency may bring many short-term successes that can be seen in cost reports but improving customer service may increase repeat customers and other factors that cannot be measured in the short-term. Both short-term and long-term successes are important to the success of the internal business perspective.

There are other possible measures that can be used to gauge the internal business perspective. In addition to those in Table 6, other measures might include: personnel spending per node or terminal; percentage of network, printer, or data store usage; cost of labor per job, page printed, or batch of jobs; ratio of database, security, or network managers per 100,000 files; and performance of manager to their budgets.

Learning and Growth Perspective: The learning and growth perspective addresses how a company will sustain and continue to create long-term growth and improvement. Technology is continually changing and companies must change with it in order to sustain their competitive advantage. In order to meet long-term goals a company must continually improve their abilities. This improvement can be measured by many employee-based measures such as employee training, employee satisfaction, and employee retention. In order for the employees to better service the business and to provide for learning and growth they must be provided with information regarding the customer and learning and growth perspective. This enables the employees to align their goals with that of the overall organization, and to make decisions based on these goals.

Table 6. Example of the internal business perspective scorecard for developing IT projects.

Goals	Measure	Actual performance	Targeted performance
Asset utilization	Percentage of CPU usage time	89%	95%
Being cost efficient	Ratio of cost of hardware (or software) investment per Mbps (*megabits per second*) rate of speed	$0.50	$0.30
Being cost efficient	Investment in IT cost per terminal	$250	$220
Higher quality	Number of errors per month, per operator	3	0

To be effective, the measures contained in these perspectives should be accurate, objective, and verifiable. Malina and Salto (2001) suggest that if the measures do not contain these qualities, a company can have trouble with managers in bad faith manipulation of the measures or managers in good faith achieving the measures and yet, causing harm to the company.

Table 7. Example of the learning and growth perspective scorecard for developing IT projects.

Goals	Measures	Actual performance	Targeted performance
Learning new systems	Average time required to fully master new IT	28 days	20 days
Implement new systems	Average time required to fully implement new IT	95 days	75 days
Enhance innovation	Number of experiments with new IT per year	3	10

An important factor that ties these scorecards together and makes the balancing process a critical factor for success is the reality that some of the measures used in the balanced scorecard method are "leading" indicators for change and some or "lagging" indicators. All of the measures are used as "indicators" of change, but some are the results of others. The financial results or the financial perspective measures should be viewed as *lagging indicators*, since they reflect the final results of making changes brought on by the actions of the *leading indicators* (i.e., customer, internal business, and learning and growth perspectives). This is why in Figure 3 we have arrows connecting all the scorecards and their bases in strategic planning. Kaplan and Norton (1996a) view the four perspectives on a continuum much like the one in Figure 4, where the double pointed arrow represents the continuum. The example of how training leads eventually to more sales or profit illustrates what is called the *causal chain* of the four perspectives of the balanced scorecard and shows how they are interrelated. Since profits can also provide more funds for training, this causal chain can work down the continuum (opposite direction) as well.

Illustration of the balanced scorecard method

To illustrate the implementation of a balanced scorecard, let's look at a hypothetical case situation to show the establishment of a balanced scorecard and how it can be used to address important business issues. A new general manager (GM) of an outsourcing IT facility realized that something radical needed to be done at the facility. When hired as the GM several months ago, the facility was in great condition, the sales figures were terrific but recently things had begun to slide downward.

One of the problems observed was that the GM was beginning to spend more and more time on employee absences and resignations and that the schedule was overscheduled most days to make up for the employees who would not show up. This was causing labor costs to increase. Overtime pay for the employees who were replacing the no-shows was also climbing. Another concern was increased competition from competitors who where opening their businesses near the outsourcing facility where the GM worked.

A closer examination of the problems revealed that the high sales the facility had experienced in the past were do in large part to large advertising expenditures and increases in other costs as well. The facility was also facing a large amount of employee and customer complaints. The employee complaints were based on the low-wages and working-hours conflicting with school activities (many of the part-time tech people were young high-school or early college aged and needed to spend time at school). The customer complaints stemmed from poor service. The facility, not providing the desired IT (i.e., a result in poor investments), had slow customer order processing. The customers that frequented the facility stopped by several times a week to acquire a wide-range of outsourcing services and were looking for full service convenience. Other customers were more price conscious in their purchases.

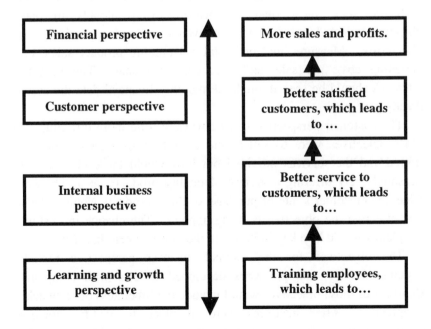

Figure 4. Leading and lagging indicator continuum in the balanced scorecard method and example.

When the quarterly financial report was released the financial figures were lower than they had been in the past. This raised concerns about the new GM's capabilities from the people who had hired the GM.

To deal with this situation the GM decided to use a balanced scorecard approach. To begin, the GM first needed to decide what the strategic goals of the firm were from the company's mission statement. Two of the strategic goals the GM derived from the mission statement and the prior strategic plans were: convenience and low-cost outsourcing products. If implemented correctly, the GM felt the customers would come, because it was convenient at a good value, the employees would enjoy coming to work, and the interaction between customers and employees would provide a good experience for both parties. This would solve many employee problems and the enhanced positive attitude of the employees would encourage customers to become repeat shoppers.

Once strategic goals were decided, the GM next needed to create objectives and measures for each of the four perspectives within the balanced scorecard. The financial perspective seemed the logical place to begin, since most financial measures were already in place. Objectives selected included increase revenue, and reduce marketing and labor costs. Measures included total outsourcing product sales, total marketing costs per sale, and total costs per sale. The measured outcomes should be seen through the monthly financial data the system already generated.

The customer perspective is, as always, a little more difficult. The GM's objectives were to satisfy the customers through the products offered and the services provided but how would these objectives be measured? By offering a variety of quality products, at a substantial value, the customer's rate of purchase of those products would measure how effective the objective is being applied. The change in customer complaints would be a key indicator to the improvement in service.

The objectives for the internal business perspective included customer service, and processes developed to deal with complaints. Measures of those objectives would include the timeliness of feedback to complaints. After measures were in place to rectify employee complaints, the staff would become more focused, evidenced by less absenteeism and a higher rate of retention. The improvement in employee attitude without the balanced scorecard would not be measured under a traditional system. The benefits of improved employee moral will benefit the business in both the short-term through the indicator such as a lower rate of absenteeism and also in the long-term through a greater retention rate.

The learning and growth perspective of the balanced scorecard included objectives such as increased customer order accuracy and quicker customer order checkouts. The GM decided that staff training would be the best method for improving these factors and implemented programs to get the staff more comfortable with processing customer orders and other business transactions. The computer error log would serve as a measure for evidence of the results of this training. Better training also would encourage customers (i.e., they would have greater confidence in the service provider) and make the staff more confident in their work.

While the ideas proposed by the GM in this hypothetical problem above seem like just good management practices of the past, they all include an important current concept of "measurement", that the balanced scorecard brings to process of management. Once set up, this balanced scorecard would provide monthly measures of performance in all four perspectives, allowing the GM in this case to monitor, adjust and change strategies in an effort to continually fine-tune them for maximum outcome.

In this sense, the balanced scorecard acts as a means of continuous improvement to enhance goal accomplishment as both a short-term and long-term planning function.

Final word on balanced scorecards

Because the balanced scorecard methodology can be used at any level in the organization, it has many opportunities for use in IT investment planning. Zee (2002, pp. 60-92) has suggested it can be used with CSFs to examine costs of IT investments, including maintenance costs of MIS, IT infrastructure, IT research, and development of new IT applications. Others have suggested that it can be used for planning enterprise systems

(Rosemann, 2001), strategic planning of IT e-commerce decisions (Raisinghani, 2001), leveraging the value delivered by IT (Meyerson, 2001), and software decisions (Eickelmann, 2001).

The advantages of using balanced scorecards can be summarized to include:

1. *Prevention of sub-optimization by the organization*: This methodology forces managers to consider all operational measures together, subjective and objective measures.   It motivates managers to think in terms of the whole organization and how their areas of responsibility fit into the whole. As the balanced scorecards are used over time, managers learn of their inter-related roles and how one department or one division supports others and their collective use of an organization's IT resources.

2. *Coordinates the needs of all stakeholders*: With every department in an organization requiring IT investments to support their customers, suppliers, and staff, there is a critical need to coordinate the timing of investment activities.   For example, launch dates to install new versions of software organization-wide must be planned to be done within the limitations of available computer staff and financial capabilities. At the same time such an undertaking must also be coordinated within a firm to best service the stakeholders.   The balanced scorecard can act to coordinate the financial, customer, and internal perspectives for maximum results.

3. *Helps educate and increase the use of IT investment methodology as a means of measuring business performance*: As managers use the balanced scorecard and the various measures, many of which are financial in nature, it helps to train in the use and appreciation of all the financial and non-financial measures organizations use in IT investment decision-making.   It helps motivate staff to think about what they do and helps them to focus on activities that will produce results in the form of the measures in the balanced scorecard.

Summary

This chapter presented three conceptual methodologies for use in IT planning: CSFs, Delphi method, and the balanced scorecard method. CSFs and the balanced scorecard methods were presented as strategic planning methods that help to tie an organization's strategic goals to all areas within the organization. The Delphi method was presented as an adjunct decision aid useful in assisting the identification of CSFs and useful as a stand-alone methodology where conceptual decision-making is required. These three methodologies where presented together in this chapter because they overlap each other in their conceptual nature in one very important decision-making aspect: they all can be the first step in IT investment decision-making. As all decision scientists know, the first step is always the most difficult. The methodologies presented in this chapter help to tie that first step of IT planning to the central core planning of all businesses (i.e., strategic plans of the business). By building on the strategic plan, all IT investment decisions are in a better position for justification, and justification for investment is the central issue in being permitted to launch an IT investment acquisition process.

CSFs, the Delphi method, and the balanced scorecard method are all examples of conceptual methods that utilize multi-criteria in their assessments. In the next chapter we present methodologies that quantify multi-criteria, which in turn could be used to support the conceptual methods presented in this chapter.

Review Terms

Balanced scorecard method
Causal chain
Cost drivers
Critical success factors (CSFs)
Delphi method
General manager (GM)
Information technology (IT)

Lagging indicators
Leading indicators
Management information systems (MIS)
Megabits per second (Mbps)
Mission statement
Value drivers

Discussion Questions

1. How many CSFs do you think a firm can have?
2. What would be the difference between a CSF at the strategic level of the organization and the operational level of the organization?
3. Why is choosing the experts when using the Delphi method critically important? Why not just use staffers?
4. Can the Delphi method be used in the balanced scorecard method?
5. What is meant by the term "perspectives" in the balanced scorecard method?
6. Why should the four "perspectives" of the balanced scorecard method be viewed as being on a continuum?

Concept Questions

1. Where do we go to determine CSFs? Explain.
2. How does the Delphi method help in determining CSFs?
3. How would you design the steps in a Delphi method process to determine advances in IT?
4. What are four of the advantages of the Delphi method?
5. What are three strengths and three weaknesses of the Delphi method?
6. What are the four perspectives of the balanced scorecard method? Explain each.
7. Since both CSFs and the balanced scorecard method use "value drivers" to improve organization performance, what is the difference between the two methodologies?
8. What is the difference between "leading indicators" and "lagging indicators" in the balanced scorecard method?
9. Why should the use of the balanced scorecard method begin with an organization's strategic plan?
10. What are some of the advantages of the balanced scorecard method?

References

Digman, L.A., *Strategic Management*, 2nd ed., Homewood, IL: BPI/Irwin, 1990.

Kaplan, R.S. and Norton, D. P., "Linking the Balanced Scorecard to Strategy," *California Management Review*, July, 1996b, pp. 53-79.

Kaplan, R.S. and Norton, D.P., "Putting the Balanced Scorecard to Work," *Harvard Business Review*, September-October, 1993, pp. 134-142.

Kaplan, R.S. and Norton, D.P., "The Balanced Scorecard-Measures That Drive Performance," *Harvard Business Review*, January-February, 1992, pp. 71-79.

Kaplan, R.S. and Norton, D.P., "Using the Balanced Scorecard as a Strategic Management System," *Harvard Business Review*, January-February, 1996a, pp. 75-85.

Laudon, K.C. and Laudon, J.P., *Management Information Systems: Managing the Digital Firm*, 8th ed., Upper Saddle River, NJ: Prentice Hall, 2004.

Malina, M.A. and Salto, F.H., "Communicating and Controlling Strategy,' *Journal of Management Accounting Research*, Vol. 35, 2001, pp. 48-90.

Martinsons, M., Davison, R. and Tse, D., "The Balanced Scorecard: A Foundation for the Strategic Management of Information Systems," *Decision Support Systems*, Vol. 25, 1999, pp. 71-88.

Meyerson, B., "Using a Balanced Scorecard Framework to Leverage the Value Delivered by IS," in Grembergen, W. V., *Information Technology Evaluation Methods & Management*, Hershey, PA: Idea Group Publishing, 2001, pp. 212-230.

Raisinghani, M., "A Balanced Analytic Approach to Strategic Electronic Commerce Decisions: A Framework of the Evaluation Method," in Grembergen, W. V., *Information Technology Evaluation Methods & Management*, Hershey, PA: Idea Group Publishing, 2001, pp. 185-197.

Rosemann, M., "Evaluating the Management of Enterprise Systems with the Balanced Scorecard," in Grembergen, W. V., *Information Technology Evaluation Methods & Management*, Hershey, PA: Idea Group Publishing, 2001, pp. 171-184.

Wen, H.J., Yen, D. and Lin, B., "Methods for Measuring Information Technology Investment Payoff," *Human Systems Management*, Vol. 17, No. 2, 1998, pp. 145-155.

Young, S.D. and O'Byrne, S.F., *EVA and Value-Based Management*, New York, NY: McGraw-Hill, 2001.

Zee, H.V.D., *Measuring the Value of Information Technology*, Hershey, PA: Idea Group Publishing, 2002.

Chapter 8

Multi-Factor Scoring Methods and the Analytic Hierarchy Process

Learning Objectives

After completing this chapter, you should be able to:

- Define and describe two types of multi-factor scoring methods.
- Use multi-factor scoring methods to make IT investment decision choices.
- Describe how the analytic hierarchy process can be used to make IT investment decision choices.
- Understand and use the analytic hierarchy process methodology to generate priorities useful in IT decision choices.
- Understand and use consistency statistics to support the analytic hierarchy process results.
- Understand how spreadsheets can be used to model multi-factor decision-making and analytic hierarchy process problems.

Introduction

IT investment decisions can be very complex if a wide range of differing factors (or criteria) are used in the decision-making process. For example, purchasing a single PC within an integrated computer system requires consideration of many factors. The selection of one manufacturer's PC over another manufacturer's PC in a network for a university student computer center can include consideration of factors such as purchase price, compatibility with other existing computers, software systems compatibility issues, the computer's features, the manufacturer's brand name, the technical support availability, historic cost of repairs, warrantee support, and flexibility in features for future adaptation with other systems. Moreover, the complexity caused by the number of factors to consider is increased by the differing nature of how the factors will be measured. While "price" and "cost" factors can be easily measured objectively in dollars for a comparison, factors such as "flexibility", "brand name", and "compatibility" have to subjectively be rated by some type of score (i.e., a "1" representing "poor" score up to a "9" for a "good" score). Still other factors, like "features" can only be counted for comparison purposes. Since each of these factors, and many more, might be important in an IT investment decision some means has to be found to bring all of these factors used as criteria in the decision-making process into a common unit of measurement. This common unit of measurement is a subjective rating system that converts objective and subjective factor measures into *scores*. A group of methodologies that make use of these subjectively derived scores when applied to selection-type decision-making problems with differing factors, are called *multi-factor scoring methods* (MFSM's). We will examine two basic types of MFSM's in this chapter. In addition, one of the many multi-factor scoring methodologies used to bring greater objectivity to an otherwise subjective process is called the *analytic hierarchy process* (AHP). We will examine how AHP is used to establish mathematical weighting used in MFSM's and how this methodology provides a mathematical weighting process to permit a more precise representation in the computations of the scores and in a final decision process. Adler (2000) considers these "new methods" for strategic IT investment decision-making.

What are Multi-Factor Scoring Methods?

Multi-factor scoring methods (MFSM) are a collection of quantitative methodologies that can be used to make a choice from a set of alternatives using a set of two or more factors as decision choice criteria

(Renkema and Berghout, 1997). The alternatives must be mutually exclusive and discrete choices (i.e., no proportional choices of more than one alternative) like those presented in Table 1 (i.e., Alternatives A, B, or C computer systems). Selecting between differing software applications to do the same task or differing manufacturer's computer systems are examples of mutually exclusive and discrete choice alternatives in IT investment decision-making. The factors used as decision choice criteria must also be rated in some numerical fashion. Any numbered rating scale such as 1 to 9, or 1 to 100, etc. can be created and used in MFSM's. In Table 1, the three alternatives have been rated on a scale of 1 (i.e., a "1" represents a "poor rating" in satisfying that criteria) to 9 (i.e., a "9" represents a "good rating" in satisfying that criteria). We can see in Table 1 that for the decision factor of "flexibility" Alternative C is rated the highest and Alternative B is rated the lowest.

Table 1. Multi-factor scoring method table for un-weighted problem.

Factors (Criteria)	Alternative A Computer System	Alternative B Computer System	Alternative C Computer System
1. Flexibility	5	1	9
2. Brand name	5	7	2
3. Price	6	3	6
4. Delivery	5	6	7
Total Score	21	17	24

Types of multi-factor scoring methods

There are many different versions of MFSM that exist in management decision-making literature (Render and Stair 2000). They can be generally categorized into two basic types: un-weighted MFSM and weighted MFSM. In *un-weighted MFSM* problems the ratings are simply summed up to achieve a score that will denote the desired choice of alternatives. For example, in Table 1, the Total Score row (note the darkened values) represents a summation of the four column scores for each of the factors or criteria being used in the selection process. Since the larger score denotes the "best" score, the un-weighted MFSM choice would be Alternative C with a score of 24.

The selection of Alternative C in Table 1 based on the un-weighted MFSM score implies that each of the four factors in the decision process were equally weighted in the mathematical process. It is more likely in real-world problems that the factors will have differing weights because of unique organizational or system requirements on IT. For example, when a company is cash-short and needs a computer system they will be inclined to wait a factor such as "price" over "brand name". Determining a *factor weight* (i.e., a mathematical weight reflecting the importance of that factor relative to the other factors) can be done very subjectively or more objectively with other quantitative methods (as we will see later in this chapter when we discuss the Analytic Hierarchy Process). For illustration purposes lets assume we are facing the same three alternative/four factor problem in Table 1, but now a factor weight for each of the four factors has been assigned as shown in Table 2. Lets say the origin of the factor weight in Table 2 is subjectively "guessed-at" by an information systems manager who feels that the factor of "flexibility" in a computer system is five times as important as the factor "brand name" and two and a half times as important as the "price" and "delivery" factors. These factor weights are usually expressed as decimals (or percentages) and they must add up to 1.0 (or 100 percent) over all the factors being considered in the problem. Note we will identify tabled computations with darkened values from this point on in the chapter.

Taking the factor weights and multiplying them times each of their related alternative ratings in each row as presented in Table 2, modifies the ratings of the factor's and alternative's to reflect their proportioned importance. These are the darkened values in the factor rows of Table 3. We next sum them up by column for each alternative in Table 3 to result in a total score value which again is used for the final alternative choice of computer system based on the largest total score. In this weighted

MFSM example, we again would select the Alternative C computer system since its total score of 7.3 is larger than the other two alternatives.

Table 2. Multi-factor scoring method table for weighted problem.

Factors (Criteria)	Factor Weight	Alternative A Computer System	Alternative B Computer System	Alternative C Computer System
1. Flexibility	0.5	5	1	9
2. Brand name	0.1	5	7	2
3. Price	0.2	6	3	6
4. Delivery	0.2	5	6	7
Total Score	1.0			

Table 3. Computation for the multi-factor scoring method weighted problem.

Factors (Criteria)	Factor Weight	Computations for Alternative A Computer System	Computations for Alternative B Computer System	Computations for Alternative C Computer System
1. Flexibility	0.5	0.5 x 5 = **2.5**	0.5 x 1 = **0.5**	0.5 x 9 = **4.5**
2. Brand name	0.1	0.1 x 5 = **0.5**	0.1 x 7 = **0.7**	0.1 x 2 = **0.2**
3. Price	0.2	0.2 x 6 = **1.2**	0.2 x 3 = **0.6**	0.2 x 6 = **1.2**
4. Delivery	0.2	0.2 x 5 = **1.0**	0.2 x 6 = **1.2**	0.2 x 7 = **1.4**
Total Score	1.0	5.2	3.0	7.3

Summary of multi-factor scoring methods solution procedures

In summary, the procedure for un-weighted MFSM includes the following steps:

1. Identify all alternative choices.
2. Identify all relevant factors.
3. Construct a MFSM table with individual columns for each alternative, rows for each factors, and final row labeled Total Score.
4. Rate each alternative using a scale of choice (e.g., 1 to 9), where the lower value on the scale represents a less preferred value and the higher value represents a more preferred value for each factor.
5. Place the ratings by row and column in each cell that makes up the table.
6. Sum the ratings in each column (i.e., each alternative) to generate a total score and place these values in the Total Score row at the bottom of the table.
7. Select the alternative with the largest total score.

In summary, the procedure for weighted MFSM includes the following steps:

1. Identify all alternative choices.
2. Identify all relevant factors.
3. Identify, judgmentally derive, or compute factor weights for each factor.
4. Construct an MFSM table with individual columns for each alternative and one additional column labeled Factor Weights, rows for each factor, and final row labeled Total Score.
5. Rate each alternative using a scale of choice (e.g., 1 to 9), where the lower value on the scale represents a less preferred value and the higher value represents a more preferred value for each factor.
6. Place the ratings by row and column in each cell that makes up the table.
7. Place the factors weights in the Factor Weight column.
8. Multiply the factors weights in the Factor Weight column times each of the ratings across each row, and place those values in each of the rating cells of the table.

9. Sum these computed ratings by column (i.e., each alternative) to generate a total score and place these values in the Total Score row at the bottom of the table.
10. Select the alternative with the largest total score.

Sensitivity analysis of multi-factor scoring methods

Both the un-weighted and weighted MFSM's are fairly simple to understand and use. Unfortunately the simplicity of following their step-wise procedures absolutely can lead to less than desirable results in IT investment decision-making. In situations where the difference between alternative choices is very small (i.e., say two alternatives with total scores of 10 and 10.0001) MFSM's procedures might lead to a choice that is in error because of the subjective nature of the measures used in those quantitative procedures. The potential for estimation error in parameters that are subjectively derived is very great and can lead to error in the final decision choice. For example, could the information systems manager who guessed at the factor weighting feel comfortable with a factor weight of 0.51 instead of 0.50 in the problem in Table 3? What about 0.52, or 0.53, etc.? The point is, when subjective measures are used to calculate exact numbers, there needs to be some additional analysis performed to assure the decision makers that the subjectively derived factor weights or the ratings themselves are not so sensitive that a small, highly probable change in one parameter in the model might change the entire decision choice solution. This additional analysis is often referred to as *sensitivity analysis*. While there are different types of sensitivity analyses (some will be discussed in other chapters) we will limit our discussion here to a simple parameter simulation exercise.

To check and see how sensitive a solution is to a change in a single parameter used in a model, a series of "what-if" scenarios are implemented. This implementation takes the form of repeatedly changing a suspect parameter with incremental, alternative values and re-computing the decision outcomes recommendations based on the quantitative results of the model (i.e., total scores). Ideally, all the parameters that are subjectively derived should be checked using this "what-if" method, but clearly some parameters used in models are more suspect of being in error than others. For each suspect parameter that the incremental changes are made, they must be made one-at-a-time (e.g., like a rating of 10 being set a 10.1, then 10.2, 10.3, etc.) holding the other parameters constant until an absolute threshold value (i.e., a maximum range of change the parameter can possibly be in error or can actually change) is achieved without an impact on the existing solution or until the solution changes. If the solution changes, then it is a sign that the parameter is very sensitive and that estimation error is likely to lead to an error in decision. It also means that additional effort is necessary to make the parameter more accurate and less error-ridden. Maybe additional data collection or expert judgment is necessary to reduce the possibility of a parameter possessing the necessary variation to cause the model's existing solution to become invalid. If the solution does not change within the maximum possible range of change for that parameter, then the existing solution can still be considered acceptable and the analysis helps to lend creditability to the previous decision. In either case, sensitivity analysis can help to make better IT investment decisions.

To illustrate this process, lets assume the "flexibility" factor weight of 0.5 can range down to 0.4 and the "brand name" factor weight can range up from 0.1 to 0.2 because of the information systems manager's personal opinion on the subject in this problem. The question is, will that much of a shift make a difference in the resulting choice of the Alternative C computer system? We could incrementally shift the values between the two factor weightings by intervals of 0.01 for ten values and re-compute all of the total scores, but to save time we simply plugged in the maximum change for the two parameters. Note, in this special case parameter situation we are simultaneously shifting two parameters at the same time. This is because the factor weights must add to one. If we were to change a single rating value, we would keep all the other parameters constant as stated in the procedure. The revised computations for this re-weighted MFSM sensitivity analysis problem are presented in Table 4. The choice in Table 4 is still the same Alternative C computer system. As such, we can feel assured that in the case that greatest possible change permitted in these two parameters will not lead to a different decision choice, and that these parameters are not very sensitive.

Table 4. Computations for revised parameter MFSM sensitivity analysis problem.

Factors (Criteria)	Factor weight	Computations for Alternative A Computer System	Computations for Alternative B Computer System	Computations for Alternative C Computer System
1. Flexibility	0.4	0.4 x 5 = **2.0**	0.4 x 1 = **0.4**	0.4 x 9 = **3.6**
2. Brand name	0.2	0.2 x 5 = **1.0**	0.2 x 7 = **1.4**	0.2 x 2 = **0.4**
3. Price	0.2	0.2 x 6 = **1.2**	0.2 x 3 = **0.6**	0.2 x 6 = **1.2**
4. Delivery	0.2	0.2 x 5 = **<u>1.0</u>**	0.2 x 6 = **<u>1.2</u>**	0.2 x 7 = **<u>1.4</u>**
Total score	1.0	5.2	3.6	6.6

Spreadsheet and computer solutions

The advantage of using a computer spreadsheet like Microsoft's ® Excel© to generate solutions is easily seen in the computational effort required in a sensitivity analysis of MFSM problems, where a single parameter is changed and all new total score values have to be computed. The Excel© formulas used is the spreadsheet generation of the answers for both the un-weighted and weighted MFSM problems (those from Tables 1 and 3) are presented in Table 5. Note columns A and B are excluded since they are used just for labeling and the factor weights as stated in previous tables. It would be easy to expand the number of rows and columns to cover any size problem. The resulting Excel© generated answers for this problem are presented in Table 6. Note that columns B, C, D, and E and rows 2, 3, 4, and 5 are the input data necessary for this program to run. All other numbers are calculated by the set of formulas in Table 5.

Table 5. Excel© printout of the Excel© formulas used in the MFSM problem.

	C	D	E	F	G	H
1	Altern. A	Altern. B	Altern. C	Altern. A	Altern. B	Altern. C
2				=B2*C2	=B2*D2	=B2*E2
3				=B3*C3	=B3*D3	=B3*E3
4				=B4*C4	=B4*D4	=B4*E4
5				=B5*C5	=B5*D5	=B5*E5
6	=SUM(C2: C5)	=SUM(D2: D5)	=SUM(E2: E5)	=SUM(F2: F5)	=SUM(G2: G5)	=SUM(H2: H5)

Table 6. Excel© printout of un-weighted and weighted MFSM problems.

	A	B	C	D	E	F	G	H
1	Factors	Factor weight	A	B	C	Altern. A	Altern. B	Altern. C
2	1	0.5	5	1	9	2.5	0.5	4.5
3	2	0.1	5	7	2	0.5	0.7	0.2
4	3	0.2	6	3	6	1.2	0.6	1.2
5	4	0.2	5	6	7	1	1.2	1.4
6	Total score	1	21	17	24	5.2	3	7.3

What is the Analytic Hierarchy Process?

In the previous weighted MFSM problem in Table 3 the factor weights were subjectively derived based on the opinion of a manager. In some problem situations such factor weights are very difficult to judgmentally estimate without considerable estimation error or in some cases they are just impossible to off-handedly guess at due to a lack of historic information. In these cases the "analytic hierarchy process" (AHP) can be employed to develop the factor weights and can also be used as a decision-making methodology in itself. Originally developed by Saaty (1980) as a decision-making technique, AHP has been applied to a wide-range of large-scale projects (Saaty and Vargus 1991) and current decision-making problems and issues (Saaty and Vargus 2001). It has also been specifically used in IT technology investment decision-making (Roper-Lowe and Sharp 1990; Schniederjans and Wilson 1991).

The analytic hierarchy process utilizes pairwise comparisons to establish factor weights for decision models, establish priorities for a decision choice, and generate accurate statistics to confirm its decision analysis. AHP is a complete decision-making process that permits a more complete consideration of multi-factors or multi-criteria than the MFSM's and as such is ideal for helping aid in the very complex and multi-factor environment of IT investment decision-making. It is a superior decision-making methodology because it requires all of the factors in the decision environment to be directly compared with all other factors, providing a more inclusive consideration of the interaction and value of each factor relative to all other factors. The workings of this methodology are best understood by learning the AHP methodological procedure.

The AHP procedure

The *AHP procedure* involves six tedious steps (broken into various sub-steps) so its presentation here will be combined with an example for illustrative purposes.

Step 1. *Establish the "decision hierarchy":* In this step the decision maker must identify: (1) the overall decision, (2) the factors that must be weighted or used to make the decision, and (3) the alternative choices from which a decision it to be made. Once these are identified they are placed in a *decision hierarchy* similar to Figure 1. The idea of the decision hierarchy is to establish the relationships of the decision, its factors, and alternatives in a logical order or hierarchy of decision process. At the top of the decision hierarchy is the overall decision that is being faced by the decision maker. Then some "n" number of factors that leads to making the decision are placed at the next level down in the hierarchy from the overall decision. Finally, some "m" number of alternatives are placed at the next level down in the hierarchy, where "m" may be a different number for each of the differing "n" factors. Neither the "n" number of factors nor the "m" number of alternatives have to be the same. Also, the "m" number of alternatives for each of the "n" factors do not have be the same.

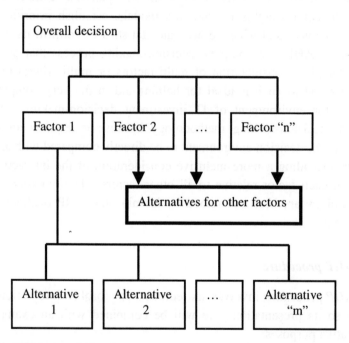

Figure 1. AHP decision hierarchy.

To illustrate the decision hierarchy construction lets apply the previous MFSM computer system selection problem. In that problem there is one overall decision (i.e., select a computer system). As shown in Figure 2, the overall decision is placed at the top of the decision hierarchy. There are also four factors used in the selection process and these are placed in the second level down from the overall decision in the hierarchy in Figure 2. Finally, the three computer alternative choices (i.e., computer system A, B, or C) are placed at the bottom of the hierarchy. In this particular problem each of the four factors has the same three alternatives below it in the hierarchy.

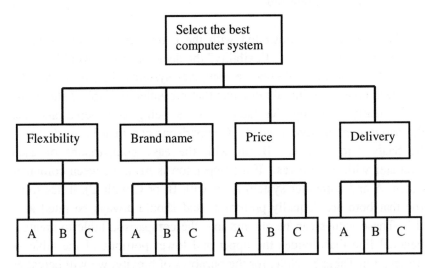

Figure 2. Decision hierarchy for computer system selection problem.

Step 2. *Establish the pairwise comparisons of alternatives*: In this step the decision maker must compare each alternative with all other alternatives, one factor at a time. This is where the term *pairwise comparison* comes from. The rating measure scale used for these comparisons forces the decision maker to chose the most desirable alternative and rate the other alternatives on a range from "equally preferred" to the most desirable alternative to "extremely preferred" as it relates to each of the factors. Note, we will also use this same scale for making comparisons between alternatives. This scale uses a 1 to 9 rating system based on Saaty (1980) original work as follows:

1. Equally preferred;
2. Equally to moderately preferred;
3. Moderately preferred;

4. Moderately to strongly preferred;
5. Strongly preferred;
6. Strongly to very strongly preferred;
7. Very strongly preferred;
8. Very to extremely strongly preferred; and
9. Extremely preferred.

This step is best illustrated with the computer selection problem. Starting with the factor of "flexibility" the decision maker would have to rate each of the three alternative computer systems (i.e., A, B, and C) with each other. An initial table like that presented in Table 7 should be used, clearly denoting the factor that is being considered and the alternative comparisons needed. Comparisons can then be made using the Saaty 1-to-9 scale. Begin with the easy comparisons of each computer with themselves. If a comparison is made between computer system A and computer system A (B with B, or C with C), it has to be true that both are "equally preferred" and should always be rated as a "1". These values can be entered into the comparison table to create a diagonal line that divides the upper and lower portions of the table as presented in Table 8. Note, for the purposes of brevity, we will hereafter refer to the "Alternative A, B, or C Computer Systems" as just the alternative of "Computer System A, B, or C."

Table 7. Initial AHP Step 2 comparison table for computer selection problem.

Flexibility	Computer System A	Computer System B	Computer System C
Computer System A			
Computer System B			
Computer System C			

Table 8. Diagonal comparisons AHP Step 2 pairwise table for computer selection problem.

Flexibility	Computer System A	Computer System B	Computer System C
Computer System A	1		
Computer System B		1	
Computer System C			1

We next need to make comparisons between differing computers. These values will go in the upper portion of the comparison table. First, we compare computer system A with B in regard to their flexibility. What we are rating is computer system B, with respect or relative to A. Note, we are rating computer B, not A. Lets say the decision maker thinks computer system B is a little more preferable (i.e., does a better or more satisfying job on providing flexibility with the existing system) than A. So the rating given is a "3" or "moderately preferred" rating to that of computer system B. When placing the rating values in the table, always start with the row, then column for the comparison. So, the rating of "3" goes in the Computer System A row, and the Computer System B column as presented in Table 9. Second, we compare computer system A with C in regard to their flexibility. Lets say the decision maker thinks computer system C is a lot more preferable (i.e., does the best job of satisfying or providing flexibility with the existing system) than A. So the rating given is a "9" or "extremely preferred" rating to that of computer system C. Again, place the rating values in the table by row, then by column. So, the rating of "9" goes in the Computer System A row, and the Computer System C column as presented in Table 9. To finish the upper portion of the table we now compare computer system B with C. Lets say the decision maker thinks computer system C rating is somewhere between the previous two ratings of "3" and "9". So the rating given is a "6" or "Strongly to very strongly preferred" to computer system C. So, the rating of "6" goes in the Computer System B row, and the Computer System C column as presented in Table 9.

Table 9. Between comparisons AHP Step 2 pairwise table for computer selection problem.

Flexibility	Computer System A	Compute System B	Computer System C
Computer System A	1	3	9
Computer System B		1	6
Computer System C			1

Now to finish the Step 2 pairwise comparisons, we need to determine the rating in the lower portion of Table 9. Since we have already made the "between" comparisons in the upper portion of the table, we can simply use the *inverse rule* to compute these values. The inverse rule simply means that the inverse of the related upper proportion values can be used for the related ratings of the lower values in the table. The logic is very simple: if we evaluated computer system B as being "3" when compared to computer system A, then we must say that computer system A is only 1/3 of a preference rating value when compared to computer system B. The cell in the Table 9 that corresponds to the comparison in row A, column B, is row B, column A. So a rating of 1/3 goes in that cell. Likewise, the cell in Table 9 that corresponds to the comparison in row A, column C which has a rating of "9", is row C, column A which should be given an inverse rating of 1/9. Finally, the cell in the Table 9 that corresponds to the comparison in row B, column C which has a rating of "6", is row C, column B which should be given an inverse rating of 1/6. The complete table showing all pairwise comparison values for the "flexibility" factor is presented in Table 10.

Table 10. Complete comparison AHP Step 2 pairwise table for computer selection problem.

Flexibility	Computer System A	Computer System B	Computer System C
Computer System A	1	3	9
Computer System B	1/3	1	6
Computer System C	1/9	1/6	1

This same process of making the comparisons must now be used for each of the other three factors. In Tables 11, 12, and 13 the stages of generating the values for the "brand name" factor are presented. Note that computer system A and computer system B were given an "equally preferred" rating of "1". This is perfectly permissible and the resulting B to A inverse value is also rated at "1".

Table 11. Diagonal comparison AHP table for brand name factor.

Brand name	Compute System A	Computer System B	Computer System C
Computer System A	1		
Computer System B		1	
Computer System C			1

For some factors, the comparison will not begin in the upper portion of the AHP comparison table, but instead begin in the lower portion as show for the "price" factor in the comparison tables in Tables 14, 15, and 16. This can happen when the upper portion comparisons require a proportion value of the scaled measures. This can happen for any factor. It is resolved by starting in the lower portion of the table for the between comparisons (as shown in Table 15) and then using the same inverse process as before to determine the upper portion values (as shown in Table 16). Note, the inverse rule does also apply to the situation here, where the values are first determined in the lower portion of the table and their inverses are used as the ratings in the upper portion.

To finally complete this step lets assume the final "delivery" factor's AHP comparison values are those given in Table 17, which happens to be the same as to those of the "brand name" factor.

Table 12. Between comparison AHP table for brand name factor.

Brand name	Computer System A	Computer System B	Computer System C
Computer System A	1	1	6
Computer System B		1	3
Computer System C			1

Table 13. Complete comparison AHP table for brand name factor.

Brand name	Computer System A	Computer System B	Computer System C
Computer System A	1	1	6
Computer System B	1	1	3
Computer System C	1/6	1/3	1

Table 14. Diagonal comparison AHP table for price factor.

Price	Computer System A	Computer System B	Computer System C
Computer System A	1		
Computer System B		1	
Computer System C			1

Table 15. Between comparison AHP table for price factor.

Price	Computer System A	Computer System B	Computer System C
Computer System A	1		
Computer System B	2	1	
Computer System C	8	5	1

Table 16. Complete comparison AHP table for price factor.

Price	Computer System A	Computer System B	Computer System C
Computer System A	1	1/2	1/8
Computer System B	2	1	1/5
Computer System C	8	5	1

Table 17. Complete comparison AHP table for delivery factor.

Delivery	Computer System A	Computer System B	Computer System C
Computer System A	1	1	6
Computer System B	1	1	3
Computer System C	1/6	1/3	1

Step 3. *Compute the factor priorities*: In this step the decision maker uses the previously determined comparison ratings to compute a set of priorities for the individual factors. To do this involves several small computation sub-steps. Begin by converting the complete comparison ratings in the tables for each of the four factors from Step 2 into decimal form as presented in Table 18. The greater the number of places behind the decimal point, the greater the precision of the resulting values. At least 4 places should be used and the last value should be rounded up for values of 5 or more. We then sum the decimal values in each column (i.e., a table for each factor but comparing the alternative computer systems) as shown by the darkened values in Table 18.

We now take the summed values from Table 18, and divide them back into the column values from which they came. The resulting ratios are the darkened values of this sub-step and are shown in Table 19. Note, summing each column is not required, but it is a useful check on the computations as these column values must equal one. If they do not, go back and re-consider the rounding so they are forced to a summation of one as shown by the darkened values in Table 19.

In the final sub-step of Step 3 we determine the priorities for the alternatives. This is accomplished by averaging the darkened ratio values in each row of Table 19. In this case we have three alternative computer systems. We take the three row values from Table 19 for computer system A row (i.e., 0.6923, 0.7200, and 0.5625) and sum them.

Table 18. Step 3 decimals and summation of all AHP tabled values.

Flexibility	Computer System A	Computer System B	Computer System C
Computer System A	1.0000	3.0000	9.0000
Computer System B	0.3333	1.0000	6.0000
Computer System C	0.1111	0.1667	1.0000
Column total	1.4444	4.1667	16.0000

Brand name	Computer System A	Computer System B	Computer System C
Computer System A	1.0000	1.0000	6.0000
Computer System B	1.0000	1.0000	3.0000
Computer System C	0.6667	0.3333	1.0000
Column total	2.6667	2.3333	10.0000

Price	Computer System A	Computer System B	Computer System C
Computer System A	1.0000	0.5000	0.1250
Computer System B	2.0000	1.0000	0.2000
Computer System C	8.0000	5.0000	1.0000
Column total	11.0000	6.5000	1.3250

Delivery	Computer System A	Computer System B	Computer System C
Computer System A	1.0000	1.0000	6.0000
Computer System B	1.0000	1.0000	3.0000
Computer System C	0.6667	0.3333	1.0000
Column total	2.6667	2.3333	10.0000

Table 19. Step 3 ratios of column total AHP tabled values.

Flexibility	Computer System A	Computer System B	Computer System C
Computer System A	$\frac{1.0000}{1.4444} = 0.6923$	$\frac{3.0000}{4.1667} = 0.7200$	$\frac{9.0000}{16.0000} = 0.5625$
Computer System B	$\frac{0.3333}{1.4444} = 0.2308$	$\frac{1.0000}{4.1667} = 0.2400$	$\frac{6.0000}{16.0000} = 0.3750$
Computer System C	$\frac{0.1111}{1.4444} = 0.0769$	$\frac{0.1667}{4.1667} = 0.0400$	$\frac{1.0000}{16.0000} = 0.0625$
Column total	1.0000	1.0000	1.0000

Brand name	Computer System A	Computer System B	Computer System C
Computer System A	$\frac{1.0000}{2.6667} = 0.3750$	$\frac{1.0000}{2.3333} = 0.4286$	$\frac{6.0000}{10.0000} = 0.6000$
Computer System B	$\frac{1.0000}{2.6667} = 0.3750$	$\frac{1.0000}{2.3333} = 0.4286$	$\frac{3.0000}{10.0000} = 0.3000$
Computer System C	$\frac{0.6667}{2.6667} = 0.2500$	$\frac{0.3333}{2.3333} = 0.1428$	$\frac{1.0000}{10.0000} = 0.1000$
Column total	1.0000	1.0000	1.0000

Price	Computer System A	Computer System B	Computer SystemC
Computer System A	$\frac{1.0000}{11.0000} = 0.0909$	$\frac{0.5000}{6.5000} = 0.0769$	$\frac{0.1250}{1.3250} = 0.0944$
Computer System B	$\frac{2.0000}{11.0000} = 0.1818$	$\frac{1.0000}{6.5000} = 0.1539$	$\frac{0.2000}{1.3250} = 0.1509$
Computer System C	$\frac{8.0000}{11.0000} = 0.7273$	$\frac{5.0000}{6.5000} = 0.7692$	$\frac{1.0000}{1.3250} = 0.7547$
Column total	1.0000	1.0000	1.0000

Delivery	Computer System A	Computer System B	Computer System C
Computer System A	$\frac{1.0000}{2.6667} = 0.3750$	$\frac{1.0000}{2.3333} = 0.4286$	$\frac{6.0000}{10.0000} = 0.6000$
Computer System B	$\frac{1.0000}{2.6667} = 0.3750$	$\frac{1.0000}{2.3333} = 0.4286$	$\frac{3.0000}{10.0000} = 0.3000$
Computer System C	$\frac{0.6667}{2.6667} = 0.2500$	$\frac{0.3333}{2.3333} = 0.1428$	$\frac{1.0000}{10.0000} = 0.1000$
Column total	1.0000	1.0000	1.0000

Next we divide the sum of the three to create the average value of 0.6583, which is a mathematical weighting that can be used as a priority ranking of the three alternatives, relative to the single factor of "flexibility". The higher the weighting, the higher the priority in the selection process. The resulting average values or the factor priorities are the darkened values shown in Table 20. Note, summing each column is not required, but it is again a useful check on the computations as these column values must equal one. If they do not, go back and re-consider the rounding so they are forced to a summation of one as shown by the darkened values in Table 20.

Based on the factor priorities computed in Table 20 we can rank the selection of the three computer systems with respect to any one factor. For example, in the case of the "flexibility" factor, we would rank or establish the priority of selection of the alternative computer systems in the following order: Select computer system A first (with a weight of 0.6583), select computer system B second (with a weight of 0.2819), and select computer system C third (with a weight of 0.0598). Again, the larger the weighting, the higher the priority in the selection process.

Step 4. *Compute the factor weight priorities*: The factor weights determined in this step are the same values used in the weighted MFSM. Here the decision maker uses the same approach as that presented in AHP procedure Steps 2 and 3, but this time it is applied to a comparison of the factors with factors. In other words, we are going to repeat the sub-steps that compared the three alternative computer systems with the same three computers systems, but this time comparing the four factors (i.e., flexibility, brand name, price, and delivery) with the same four factors. Fortunately we only have to prepare one set of these tables instead of an individual one for each of the factors and this illustration will provide a review of the sub-steps in Steps 2 and 3 again.

First, we structure the comparison table with the four factor alternatives as presented in Table 21. We next enter in the diagonal comparisons as presented in Table 22.

Table 20. Step 3 Final substep AHP priority calculations.

Flexibility	Priority calculations	Resulting priorities
Computer System A	$\dfrac{0.6923 + 0.7200 + 0.5625 =}{3}$	0.6583
Computer System B	$\dfrac{0.2308 + 0.2400 + 0.3750 =}{3}$	0.2819
Computer System C	$\dfrac{0.0769 + 0.0400 + 0.0625 =}{3}$	0.0598
Total		1.0000

Brand name		
Computer System A	$\dfrac{0.3750 + 0.4286 + 0.6000 =}{3}$	0.4679
Computer System B	$\dfrac{0.3750 + 0.4286 + 0.3000 =}{3}$	0.3679
Computer System C	$\dfrac{0.2500 + 0.1428 + 0.1000 =}{3}$	0.1642
Total		1.0000

Price		
Computer System A	$\dfrac{0.0909 + 0.0769 + 0.0944 =}{3}$	0.0874
Computer System B	$\dfrac{0.1818 + 0.1539 + 0.1509 =}{3}$	0.1622
Computer System C	$\dfrac{0.7273 + 0.7692 + 0.7547 =}{3}$	0.7504
Total		1.0000

Delivery		
Computer System A	$\dfrac{0.3750 + 0.4286 + 0.6000 =}{3}$	0.4679
Computer System B	$\dfrac{0.3750 + 0.4286 + 0.3000 =}{3}$	0.3679
Computer System C	$\dfrac{0.2500 + 0.1428 + 0.1000 =}{3}$	0.1642
Total		1.0000

Table 21. Initial AHP comparison table for factor weights.

Factor weights	Flexibility	Brand name	Price	Delivery
Flexibility				
Brand name				
Price				
Delivery				

Table 22. Diagonal comparisons AHP pairwise table for factor weights.

Factor weights	Flexibility	Brand name	Price	Delivery
Flexibility	1			
Brand name		1		
Price			1	
Delivery				1

We next have to enter the judgmentally generated ratings based on the Saaty's (1980) 1 to 9 scale. Lets say in this example, that the "flexibility" factor is equally preferred (i.e., a rating of 1) to the factor "brand name", strongly preferred (i.e., a rating of 5) to the factor "price", and very strongly preferred (i.e., a rating of 7) to the factor "delivery". Lets also say that the factor "brand name" is moderately preferred (i.e., a rating of 3) to the factors of "price" and "delivery". Lets also say that the factor "price" is equally preferred (i.e., a rating of 1) to the factor of "delivery". This information permits the between comparisons to be entered into the upper portion of the table as presented Table 23.

Table 23. Between comparisons AHP pairwise table for factor weights.

Factor weights	Flexibility	Brand name	Price	Delivery
Flexibility	1	1	5	7
Brand name		1	3	3
Price			1	1
Delivery				1

Taking the inverse of the values in the upper portion to fill the respective lower cells in the AHP comparison table results in a finished table with all comparison ratings as presented in Table 24.

Table 24. Complete comparison AHP pairwise table for factor weights.

Factor weights	Flexibility	Brand name	Price	Delivery
Flexibility	1	1	5	7
Brand name	1	1	3	3
Price	1/5	1/3	1	1
Delivery	1/7	1/3	1	1

We next convert the values in Table 24 into decimals and sum the columns as in Table 25. Continuing from Table 25, we take the column totals and divide them back into their respective individual column values as presented in Table 26.

Table 25. Decimal and summation of AHP column tabled values.

Factor weights	Flexibility	Brand name	Price	Delivery
Flexibility	1.0000	1.0000	5.0000	7.0000
Brand name	1.0000	1.0000	3.0000	3.0000
Price	0.2000	0.3333	1.0000	1.0000
Delivery	0.1429	0.3333	1.0000	1.0000
Column total	2.3429	2.6666	10.0000	12.0000

Table 26. Ratios of column total AHP tabled values.

Factor weights	Flexibility	Brand name	Price	Delivery
Flexibility	$\frac{1.0000}{2.3429}$=0.4268	$\frac{1.0000}{2.6666}$=0.3750	$\frac{5.0000}{10.0000}$=0.5000	$\frac{7.0000}{12.0000}$=0.5834
Brand name	$\frac{1.0000}{2.3429}$=0.4268	$\frac{1.0000}{2.6666}$=0.3750	$\frac{3.0000}{10.0000}$=0.3000	$\frac{3.0000}{12.0000}$=0.2500
Price	$\frac{0.2000}{2.3429}$=0.0854	$\frac{0.3333}{2.6666}$=0.1250	$\frac{1.0000}{10.0000}$=0.1000	$\frac{1.0000}{12.0000}$=0.0833
Delivery	$\frac{0.1429}{2.3429}$=0.0610	$\frac{0.3333}{2.6666}$=0.1250	$\frac{1.0000}{10.0000}$=0.1000	$\frac{1.0000}{12.0000}$=0.0833
Column total	1.0000	1.0000	1.0000	1.0000

Completing the step and generating the resulting factor weights we average the four factor values by row as presented in Table 27. These factor weights are the same as those used in the weighted MFSM. Rather than just guess at these weights, this AHP approach forces the decision maker to consider all possible comparisons between the factors, thus arriving at a more inclusive, and likely a more accurate factor weighting.

Table 27. Final AHP factor weight priority calculations.

Factor weights	Priority calculations	Resulting priorities
Flexibility	$\dfrac{0.4268 + 0.3750 + 0.5000 + 0.5834}{4} =$	0.4713
Brand name	$\dfrac{0.4268 + 0.3750 + 0.3000 + 0.2500}{4} =$	0.3380
Price	$\dfrac{0.0854 + 0.1250 + 0.1000 + 0.0833}{4} =$	0.0984
Delivery	$\dfrac{0.0610 + 0.1250 + 0.1000 + 0.0833}{4} =$	0.0923
Total		1.0000

Step 5. Compute the overall decision priorities: In this step the decision maker uses the factor weights from Step 4 (i.e., darkened values in Table 27), and the values from Step 3 (i.e., darkened values in Table 20) as they were used in the weighted MFSM procedure to compute expected values for the overall decision. In this case the decision will be determined by the calculation of *overall decision priority* weighting for each of the alternatives. These priorities can be used to make the overall decision in the decision hierarchy from Step 1. To do this, we begin by creating a table that combines the information we need for the overall decision priorities. Taking the darkened values from Table 20 and arranging them by row, then multiplying them by the darkened values in Table 27 we end up with the resulting darkened values in Table 28. Summing these darkened values in Table 28 we derive the overall decision priorities. Note that if you add the Column Totals up (0.5201 + 0.3072 + 0.1727) they once again add to one and must add to one or a computation error has taken place. This check is not required but is again recommended as a check on the accuracy of the computations.

Table 28. Overall decision priority calculations.

Overall decision priorities	Alternative	Alternative	Alternative
Factors	Computer System A	Computer System B	Computer System C
Flexibility	0.6583 x 0.4713 = **0.3102**	0.2819 x 0.4713 = **0.1329**	0.0598 x 0.4713 = **0.0282**
Brand name	0.4679 x 0.3380 = **0.1581**	0.3679 x 0.3380 = **0.1243**	0.1642 x 0.3380 = **0.0555**
Price	0.0874 x 0.0984 = **0.0086**	0.1622 x 0.0984 = **0.0160**	0.7504 x 0.0984 = **0.0738**
Delivery	0.4679 x 0.0923 = **0.0432**	0.3679 x 0.0923 = **0.0340**	0.1642 x 0.0923 = **0.0152**
Column Total	0.5201	0.3072	0.1727

The use of the overall priorities computed in Table 28 is quite simple. The larger the weight, the higher the priority in the alternative selection decision. In this computer system problem, the AHP analysis tells us the computer system A should be selected first (i.e., 1st largest weight at 0.5201), if not A, then computer system B second (i.e., 2nd largest weight at 0.3072) and finally, computer system C should be selected last (i.e., 3rd and last largest weight at 0.1727).

Step 6. *Determine consistency ratios*: As previously stated, AHP does more than just generate a solution, it includes some additional analysis which permits decision makers to investigate if the subjective ratings are consistent enough to justify using the resulting overall decision priorities. In other words, AHP checks itself to make sure the ratings consistently make sense for the purposes of using the AHP analysis on which to base a decision. There are a number of sub-steps to generate the consistency ratios. To minimize the duplication of effort and confusion on this step, only one of the four factors (i.e., the "flexibility" factor) will be used for illustrative purposes. To do a complete job of determining consistency ratios, the solution requires all four of the factor tables in Table 20 and the additional factor weights presented in Table 26.

The first sub-step is to compute the *weighted sum vector*. The values that make up the vector are found by matrix multiplication where the row values in the first matrix are multiplied down each of the columns in the second matrix and their products are added together. In Table 29 this is simply illustrated for the "flexibility" factor. We restate the decimal values for Table 18 in Table 29. We then take the resulting priorities from Table 20 for the three alternatives. They are 0.6583 for computer system A, 0.2819 for computer system B, and 0.0598 for computer system C. Note, how these three values are repeated down each column, in each row of the computations in Table 29. The darkened values in Table 29 are the desired weighted sum vector values and are related to their respective row computer system.

Table 29. Matrix multiplication computations for weighted sum vector.

	Computer System A	Computer System B	Computer System C	Weighted sum vector
Computer System A	0.6583x1.0000 +	0.2819x3.0000 +	0.0598x9.0000 =	2.0422
Computer System B	0.6583x0.3333 +	0.2819x1.0000 +	0.0598x6.0000 =	0.8601
Computer System C	0.6583x0.1111 +	0.2819x0.1667 +	0.0598x1.0000 =	0.1799

Now in the second sub-step, divide each of the weighted sum vector values by their related factor priority values previously computed in Table 20. These ratios are the desired *consistency vector* values and their computation is presented in Table 30.

Table 30. Matrix multiplication computations for weighted sum vector.

	Computations	Consistency vector
Computer System A	2.0422/0.6583 =	3.1022
Computer System B	0.8601/0.2819 =	3.0511
Computer System C	0.1799/0.0598 =	3.0084

In the third sub-step we compute the *consistency index*. This index is found using the following formula:

$$CI = \frac{\lambda - n}{n - 1}$$

were *CI* is the consistency index value, *n* is the number of items being compared (i.e., computer system A, B, and C) and λ is the average of the weighted sum vector values. In this problem *n* is equal to the three computer systems being compared and λ equal to 3.0539 (i.e., [3.1022+3.0511+3.0084]/3). So the resulting *CI* is:

$$CI = \frac{\lambda - n}{n - 1} = \frac{3.0539 - 3}{3 - 1} = 0.0270$$

The fourth and final sub-step involves computing the *consistency ratio* and interpreting it. This ratio is computed by a simple formula:

$$CR = \frac{CI}{RI}$$

where *CR* is the consistency ratio and *RI* is a "random index" value that is obtained from a computed set of tabled statistics in Table 31. The *random index* is a statistic designed to identify significant variability of statistical variation in the rating measures. The *RI* table only goes up to an *n* of ten factors or alternatives being compared, but can be used for all applications of AHP. In the case of the "flexibility" factor, we are only comparing three computer systems so the resulting *RI* is 0.52 based on Table 31.

Random

Table 31. Random index (RI) values for given "n" number of comparison.

"n" number of comparisons	1	2	3	4	5
Random index	0.00	0.00	0.52	0.89	1.11
"n" number of comparisons	6	7	8	9	10
Random index	1.25	1.35	1.40	1.45	1.49

Finally, we can compute the consistency ratio as:

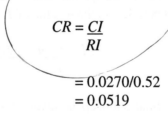

$$CR = \frac{CI}{RI}$$

$$= 0.0270/0.52$$
$$= 0.0519$$

The interpretation is that for values of CR > 0.10 there exists sufficient inconsistency that a re-evaluation of the basic factors and alternatives (that is Step 2, and all the subsequent computations in the remaining steps) should be undertaken. Simply put, there is too much inconsistency to use the AHP method and new, more carefully made comparisons are needed before a decision should be made. For values of CR ≤ 0.10 the decision maker's ratings are relatively consistent and the AHP method can be used for making a decision. As we can see in the

CR of 0.0519, there is not a sufficient amount of inconsistency to challenge the existing solution or bother going to obtain additional ratings. The other three factors and the factor weights all have CR's ≤ 0.10. This is not shown but left for students to confirm as a learning exercise in this textbook.

Summary of the AHP solution procedure

In summary, the procedure for AHP includes the following steps:

1. Establish the "decision hierarchy" by determining the overall decision, the factors and the alternatives.
2. Establish the pairwise comparisons of alternatives through a subjective judgment process and using Saaty's nine point scale.
3. Compute the factor priorities based on the values from Step 2.
4. Compute the factor weights based on the same set of procedures from Step 3.
5. Compute the overall decision priorities using a similar matrix multiplication as that of MFSM.
6. Determine consistency ratios by first computing a consistency index, and then using the random index values from the table.

Spreadsheet and computer solutions

There are a number of commercial AHP-based software applications that use spreadsheets for data entry and analysis. One of the more noted systems for the PC is the Microsoft Windows-based system called "Expert Choice" (www.ExpertChoice.com). This software will perform all of the steps presented in this chapter, plus a number of other varied and useful features that support the AHP analysis.

To illustrate how a spreadsheet can be used to perform the AHP analysis, Step 3 (which is also the same procedure for Step 4) for the "flexibility" factor is modeled in Tables 32 and 33. In Table 32, the input data required is presented. These data are the decimal values for all the factor comparisons by alternatives (i.e., the output of Step 2 for each factor). What is modeled is a template for Step 3 computations for any 3-by-3 set of alternatives problem situation. By placing, as we did in this problem the nine decimal values for the comparisons in the spreadsheet, specifically in the cells of B2 to D2, B3 to D3 and C4 to D4 for the three computer system alternatives (i.e., A, B, and C) from Table 18, the Excel© spreadsheet completed all the computation for Step 3 in the AHP procedure. Excel© first summed the column totals in cells B5 to D5, then computed the ratios that we computed in Table 19, in the Ratio A, B, and C columns. Finally, the AHP priorities that we computed in Table 20, are calculated in the H column of the Excel© matrix in Table 32. Note, there are some minor rounding differences from those values found in Tables 18, 19, and 20. The actual Excel© formulas for these calculations are presented in Table 33.

Table 32. Excel© printout of AHP Step 3 calculations.

A	B	C	D	E	F	G	H
1 Flexibility	System A	System B	System C	Ratio A	Ratio B	Ratio C	Priorities
2 System A	1	3	9	0.692329	0.719994	0.5625	0.658274
3 System B	0.3333	1	6	0.230753	0.239998	0.375	0.281917
4 System C	0.1111	0.1667	1	0.076918	0.040008	0.0625	0.059808
5 C Total	1.4444	4.1667	16	1	1	1	1

AHP is an ideal methodology for spreadsheet applications. All of the steps for AHP presented in this chapter can be placed into a single spreadsheet for easy manipulation and simulation of differing scenarios in a similar fashion to the sensitivity analysis mentioned in the section on MFSM.

Table 33. Excel© printout of formulas used in AHP Step 3 calculations.

	B	C	D
1	System A	System B	System C
2	1	3	9
3	0.3333	1	6
4	0.1111	0.1667	1
5	=SUM(B2:B4)	=SUM(C2:C4)	=SUM(D2:D4)

	E	F	G	H
1	Ratio A	Ratio B	Ratio C	Priorities
2	=B2/B5	=C2/C5	=D2/D5	=(E2+F2+G2)/3
3	=B3/B5	=C3/C5	=D3/D5	=(E3+F3+G3)/3
4	=B4/B5	=C4/C5	=D4/D5	=(E4+F4+G4)/3
5	=SUM(E2:E4)	=SUM(F2:F4)	=SUM(G2:G4)	=SUM(H2:H4)

Summary

This chapter has introduced two types of multi-factor or multi-criteria decision-making methodologies. The multi-factor scoring methods (MFSM) were presented along with their methodological procedures for both an un-weighted and weighted models. The analytic hierarchy process (AHP) and its basic methodologies were also presented as both a decision-making aid and a means of generating mathematical weights that could be used in MFSM's. AHP also included solution support statistics in the form of a consistency ratio that helps to identify where there is a need for improved comparison data necessary to insure an accurate decision. Illustrative examples were presented for each methodology along with spreadsheet applications.

Making a decision on the purchase of a computer system where subjective evaluations of decision factors like price and flexibility as illustrated in the examples in this chapter can be done very well by the multi-factor methods presented in this chapter. Unfortunately, many decisions environments face a number of limiting or constraining factors that make decision-making much more complex than the MFSM and AHP approaches can handle. When there are recognizable limitations or constraints in the decision environment, multi-factor or multi-criteria methods that are specifically designed to incorporate those constraints are necessary. In the next chapter we will learn about several multi-criteria methodologies that not only recognize real-world limitations or constraints, but also will provide decision-making choices that are optimal and can not be improved upon.

Review Terms

Analytic hierarchy process (AHP)
Consistency ratio (CR)
Consistency index (CI)
Consistency vector
Decision hierarchy
Factor weight
Inverse rule
Multi-factor scoring method (MFSM)

Overall decision priority
Pairwise comparison
Random index (RI)
Sensitivity analysis
Un-weighted multi-factor scoring method
Weighted multi-factor scoring method
Weighted sum vector

Discussion Questions

1. Where do factor weights come from in MFSM?
2. Can we solve the same type of IT problem using MFSM and AHP?
3. Do all the multi-factor methodologies in this chapter require subjective input? Explain.
4. Where do factor weights come from in AHP?
5. What step in AHP generates the same factor weights as are used in MFSM?

6. Why does the "inverse rule" used to determine ratings in AHP make sense?
7. What are each of the steps in the AHP analysis?
8. Why do you think we need a "decision hierarchy" diagram in AHP?

Concept Questions

1. What is the fundamental difference between the un-weighted and weighted multi-factor scoring methods? Explain.
2. Which of the decision-making methodologies used in IT selection in this chapter is the best? Explain.
3. What is the difference between the consistency ratio and the consistency index?
4. You have just computed a factor weight CI equal to 0.2. What do you do about that result? Describe a course of action.
5. If you have determined the overall decision priorities for two alternatives, the A alternative is 0.499 and the B alternative is 0.501, why would the size of these two priority weights justify the "consistency ratio" effort? Explain.

Problems

1. You have just computed the total scores for four data storage alternatives (i.e., DVD disks, Zip disks, CD disks, and floppies) using the weighted MFSM approach. The rating scale used was a 1 (representing an undesirable choice) to a 9 (representing a very favorable choice) for an investment in data storage technology. There are four types of data storage units being considered. The DVD technology received a total score of 5.67, the Zip disk technology received a total score of 7.88, the CD disk technology received a total score of 1.57, and floppies technology received a total score of 0.99. Which information storage alternative should be selected? Explain your answer.
2. Given the ratings for the two Web site technology applications in the table below, what are the resulting total score values for the

alternatives? Show your work. What is the best choice if the rating scale used was a 1 (representing an undesirable choice) to a 9 (representing a very favorable choice) for an investment?

Factors (Criteria)	Alternative A web site technology	Alternative B web site Technology
1. Cost of installation	5	3
2. Integration with existing software	1	9
3. Price	6	3

3. Prepare a spreadsheet answer for Problem 2 based on the spreadsheets presented in Tables 5 and 6 in the textbook. Show your answers and identify them.

4. Using the same ratings in the table in Problem 2 and the following weights, what are the resulting total score values for the alternatives, given the following weights: Cost of Installation = 0.60, Integration = 0.15, and Price=0.25? Show your work. What is the best choice if the rating scale used was a 1 (representing an undesirable choice) to a 9 (representing a very favorable choice) for an investment?

5. Prepare a spreadsheet answer for Problem 2 based on the spreadsheets presented in Tables 5 and 6 in the textbook but using the following weights: Cost of Installation = 0.40, Integration = 0.25, and Price=0.35? Show your work. What is the best choice if the rating scale used was a 1 (representing an undesirable choice) to a 9 (representing a very favorable choice) for an investment?

6. Using the same ratings in the table in Problem 2 and the following weights, what are the resulting total score values for the alternatives given the following weights: Cost of Installation = 0.20, Integration = 0.40, and Price=0.40? Show your work. What is the best choice if the rating scale used was a 1 (representing an undesirable choice) to a 9 (representing a very favorable choice) for an investment?

7. Given the ratings for the software applications in the table below, what are the resulting total score values for the alternatives? Show your work. What is the best choice if the rating scale used was a 1 (representing an undesirable choice) to a 9 (representing a very favorable choice) for an investment?

Factors (Criteria)	Alternative A Software application	Alternative B Software application	Alternative C Software application
1. Adaptability	4	3	4
2. Integration with existing software	7	1	2
3. Price	6	3	6
4. Service after the sale	5	6	3
5. Local service unit available in same city	1	1	9

8. Using the same ratings in the table in Problem 7 and the following weights, what are the resulting total score values for the alternatives, given the following weights: Adaptability = 0.20, Integration = 0.15, Price=0.12, Service=0.38, and Local Service=0.15? Show your work. What is the best choice if the rating scale used was a 1 (representing an undesirable choice) to a 9 (representing a very favorable choice) for an investment?

9. Prepare a spreadsheet answer for Problem 7 based on the spreadsheets presented in Tables 5 and 6 in the textbook. Show your answers and identify them.

10. Given n=5, what is the necessary random index value necessary to compute the consistency index?

11. Given n=3, what is the necessary random index value necessary to compute the consistency index?

12. Given n=5 and λ=5.9890, what is the resulting consistency ratio?

13. Given an n=4 and a λ=6.1230 what is the resulting consistency ratio? Comment on the consistency in this test? Prepare a specific recommendation on what should be done.

14. You must make the decision to choose one of four possible CD Read/Write manufacturer's drives (i.e., Sony, Dell, GE, and RCA) for use in your company's PC's. An evaluation of all four alternatives are used with AHP scaling methodology to determine ratings of the differing technology. You have been told to consider the following pairwise comparisons in your analysis: Sony is equally to moderately preferred to Dell, Sony is strongly preferred to GE, Sony is very strongly preferred to RCA. Dell is very strongly preferred to GE and Dell is moderately to strongly preferred to RCA. Finally, GE is moderately preferred to RCA. Given this information, prepare a comparison table required in Step 2 of the AHP procedure.

15. You have an important investment decision in selecting one type of manufacturer's brand of PC for your company to purchase for their multiple computer work centers. You must make the decision to choose one of three possible PC brands (i.e., Gateway, IBM, Dell). A very careful evaluation of all three alternatives has been undertaken with the idea of using AHP as a decision aid methodology. You have been told to consider the following pairwise comparisons in your analysis: Gateway is equally preferred to IBM, Gateway is strongly preferred to Dell. IBM is very strongly preferred to Dell. Given this information, prepare a comparison table required in Step 2 of the AHP procedure.

16. Given the comparison table below, finish Step 2 of AHP procedure and complete Step 3 to determine the factor priorities for these telecommunication alternatives as they relate to the factor of "price".

Price	Internet System	Electronic Data Interchange System (EDI)	Fax System
Internet System	1	9	7
EDI System		1	5
Fax System			1

17. Given the comparison table below, finish Step 2 of AHP procedure and complete Step 3 to determine the factor priorities for these software alternatives as they relate to the factor of "Integration".

Integration	WORD	COREL	ROXIE	PAPERPORT
WORD software	1			
COREL software	2	1		
ROXIE software	3	9	1	
PAPERPORT software	5	6	1	1

18. Given the comparison table below, perform Step 4 of AHP procedure and determine the factor weights for these three factors.

	Ease with which the IT can be repaired	Ease of use	Available training	Available technical support
Ease with which the IT can be repaired	1			
Ease of use	2	1		
Available training	7	9	1	
Available technical support	2	2	1	1

19. The tables below represent two factors (i.e., Price and Quality) and two alternatives   (i.e., Outsource or Not) in an AHP problem. The overall decision is to decide which of the two alternatives is the best.   Perform all six steps of the AHP procedure to arrive at an overall decision. Comment on the CR's but do not revised your solution.

Price	Outsource information system work	Do not use outsourcing, do the work in-house
Outsource information system work	1	
Do not use outsourcing, do the work in-house	9	1

Price	Outsource information system work	Do not use outsourcing, do the work in-house
Outsource information system work	1	4
Do not use outsourcing, do the work in-house		1

Factor weights	Price	Quality
Price	1	3
Quality		1

20. The tables below represent three factors (i.e., Cost, Flexibility, and Integration) and three alternatives (i.e., System A, System B, or System C) in an AHP problem. The overall decision is to decide which one of the three alternatives is the best. Perform all six steps of the AHP procedure to arrive at an overall decision. Be sure to include the decision hierarchy diagram and show your analysis at all six steps. Comment on the CR's but do not revise your solution.

Cost	System A	System B	System C
System A			
System B	9		
System C	6	7	

Flexibility	System A	System B	System C
System A		3	2
System B			9
System C			

Integration	System A	System B	System C
System A		6	4
System B			1
System C			

Factor weights	Cost	Flexibility	Integration
Cost		3	6
Flexibility			5
Integration			

References

Adler, R.W., Strategic Investment Decision Appraisal Techniques: The Old and New," *Business Horizons*, November-December, 2000, pp. 15-22.

Renkema, T.J. and Berghout, E.W., "Methodologies for Information Systems Investment Evaluation at the Proposal Stage: A Comparative Review," *Information and Software Technlogy*, Vol. 39, 1997, pp. 1-13.

Render, B. and Stair, R.M., *Quantitative Analysis for Management*, Supplement 11, Upper Saddle River, NJ: Prentice Hall, 2000,.

Roper-Lowe, G.C. and Sharp, J.A. "The Analytic Hierarchy Process and Its Application to n Information Technology Decision", *Journal of Operational Research Society*, Vol. 41, 1990, pp. 49-59.

Saaty, T.L., *The Analytic Hierarchy Process*, New York, McGraw-Hill Book Company, 1980.

Saaty, T.L. and Vargas, L.G., *Models, Methods, Concepts & Applications of the Analytic Hierarchy Process*, Norwell, MA: Kluwer Academic Publishers, 2001.

Saaty, T.L. and Vargas, L.G., *Prediction, Projection and Forecasting*, Norwell, MA: Kluwer Academic Publishers, 2001.

Schniederjans, M.J. and Wilson, R., "Using the Analytic Hierarchy Process and Goal Programming for Information System Project Selection," *Information and Management*, Vol. 20, 1991, pp. 333-342.

Chapter 9

Decision Analysis and Multi-Objective Programming Methods

Learning Objectives

After completing this chapter, you should be able to:

- Explain what "decision analysis" is and how it can be used in IT investment decision-making.
- Explain how to compute answers for a variety of "decision analysis" methods.
- Define the decision environments under which "decision analysis" can operate, and explain how differing environments require differing methodology.
- Explain what "goal programming" is and how as a multi-objective programming approach can be used in IT investment decision-making.
- Understand how to formulate "goal programming" models.

Introduction

This chapter describes two commonly used sets of decision-making methodologies: decision theory and multi-objective programming. *Decision theory* is really a collection of methodologies and principles used to make single, alternative choice decisions (i.e., where the decision is to select one alternative from a set of others). The procedural mathematics used to render a decision using decision theory methods and their application in IT decision-making will be presented.

233

Multi-objective programming (MOP) methods are also a collection of decision-making methods, capable of choosing any number of alternatives from a set of alternatives or proportions of alternatives where the problem requires that type of solution. Due to the complexity o f MOP methods, our discussion in this chapter will be limited to one type of MOP (i.e., goal programming) and only the formulation and solution interpretation.

What is Decision Theory?

Decision theory (DT) is a field of study that applies mathematical and statistical methodologies to help provide information on which decisions can be made (Meredith *et al.* 2002, pp. 221-269; Moore and Weatherford, 2001, pp. 399-441; Savage, 2003, pp. 183-220). Before we can use these DT methodologies we must know the elements of the DT model so as to identify and correctly formulate the problem. Note, that the use of the words "problem formulation" and "model formulation" to mean the same thing, so we will simply refer to them as "problem/model formulation."

Decision theory problem/model elements

There are three primary elements in all DT problems: alternatives, states of nature, and payoffs:

1. *Alternatives*: (sometimes called "choices" or "strategies") are the independent decision variables in the DT model. They represent the alternative strategies or choices of action that you select from. When only one choice is allowed, it is called a pure choice problem. We will limit our discussion to pure choice DT problems.

2. *States of Nature*: are independent events that are assumed to occur in the future. For example, in economics recession, depression, or growth periods are considered states of nature. In

horse racing to win, place, show, or lose are the four states of nature a horse can experience in a race.

3. *Payoffs:* are dependent parameters that are assumed to occur given a particular alternative is selected and a particular state of nature occurs. Payoff values may be in terms of profit or cost. They may be stated using positive numbers, negative numbers, or with a zero.

We combine these three primary elements into a *payoff table* to formulate the DT problem/model. The general statement of a DT problem/model is presented in Table 1.

Table 1. Generalized statement of the DT problem/model.

| Alternatives | States | of | Nature |
	1	2	...	n
1	P_{11}	P_{12}	...	P_{1n}
2	P_{21}	P_{22}	...	P_{2n}
:	:	:	:	:
m	P_{m1}	P_{m2}	...	P_{mn}

In the generalized statement of the DT problem/model we can have m alternatives and n states of nature. The idea here is that there can be a different number of alternatives than states of nature (i.e., so m does not have to equal n). Also, the P_{ij} (where $i=1, 2,..., m$; $j=1,2, ..., n$) payoff values are listed by row and column denoting that if a particular alternative is selected and a particular state of nature occurs, the decision-making will be rewarded with the specific P_{ij} payoff. The alternatives are always listed in DT problem formulations as rows and the states of nature always as columns.

DT problems are pure choice problems, which are very applicable to the types of problems faced in IT investment decision-making. What complicates the problem is that the exact payoff is dependent on the nature of the decision environment the decision maker is facing.

Types of decision environments

There are three primary types of DT environments managers' face: certainty, risk, and uncertainty:

1. *Certainty*: Under this environment the decision maker knows clearly what the alternatives are to choose from and the payoffs that each choice will bring with certainty if the alternative is chosen. For example, if you go into a computer store and buy a computer based only on the size of its storage space, the computer storage space information is usually available and known with certainty before purchase. The computer space is not open to judgment or estimation, it is known with certainty based on tested technology.

2. *Risk*: Under this environment some information on the payoffs are available but are presented in a probabilistic fashion. For example, if you want to invest in one of two technologies, and one has a 50 percent chance of being a successful investment (or a 50 percent chance of failure) and the other has a 30 percent chance of success (or a 70 percent chance of failure), this would be a risk situation. It is risky because you only know partially or on a percentage basis what the states of nature (i.e., success or failure) will be.

3. *Uncertainty*: Under this environment no information about the likelihood of states of nature occurring is available. We can only assume that a particular payoff will occur if a given state of nature occurs. For example, the payoffs of a new IT in a "good" or "bad" decision situation might be estimated, but are only assumed to occur in an uncertain environment. They are not known with any degree of certainty.

You can look at the three environments as being on a linear continuum ranging from complete knowledge (i.e., under certainty), to partial knowledge (i.e., under risk), and finally to no knowledge of the states of nature occurring (i.e., under uncertainty). With each of these environments, there are different criteria that one can use to aid in making a decision. Moreover, each environment has a variety of methodologies to derive payoff information on which a decision can be made.

Decision Theory Formulation and Solution Methodologies

In this section we will first see how to formulate the DT problem/model and then solve it. The means by which we solve a DT problem depends on the type of decision environment we face. We will examine a number of approaches suggested for solving DT problems covering all three decision environments.

A decision theory problem/model formulation procedure

The procedure for formulation of a DT problem/model consists of the following general steps:

 1. *Identify and list as rows the alternatives to choose from.*
 2. *Identify and list as columns the states of nature that can occur.*
 3. *Identify and list in the appropriate row and column the payoffs.*
 4. *Formulate the problem/model as a payoff table.*

Using this procedure, let's look at two IT investment problems.

A Network Topology Problem

Suppose an information systems manager wants to decide which of two types of computer networks to configure to serve customers. The choice in networks is either a ring network (i.e., in which all the computers are linked by a closed loop in a manner that passes data in one direction from computer to computer) or a star network (i.e., in which all the computers and other IT are connected to a central host computer so that communications must pass through the host computer). These two network configurations are the alternatives in this problem. What the manager will receive (i.e., the payoffs) depends on the customer demand conditions (i.e., states of nature) that define the service levels provided by the network chosen. The two possible customer demand conditions in this problem are a "High Demand" or a "Low Demand." If manager uses a the ring network and experiences a "High Demand" condition, the network will be able to process customer requests such that it can generate $3 million in sales a day. If the ring network experiences a "Low Demand" condition, it will only be able to process sales equal to $1 million per day. If the manager chooses the star network and experiences a "High Demand" condition, the network will be able to process customer requests generating $4 million in sales. If the network experiences a "Low Demand" condition, the network would actually incur a loss equal to the firm of $2 million in sales. Let us further state that the manager has contractual agreements with the customers in this problem that guarantee the payoffs as being certain given our choice of alternatives. What is the DT problem/model formulation for this problem?

Because there is "certainty" in payoffs in this problem we can treat it as a "certainty" type DT problem. It should be mentioned that when we finish formulating the problem, we ascertain which environment this problem will be from by just looking at its formulation. As stated above in the problem, the type of environment must be defined. Using the four-step DT procedure we can formulate this problem/model accordingly:

1. *Identify and list as rows the alternatives to choose from.* There are two alternatives: Ring Network and Star Network. Only one will be chosen, thus a pure choice problem.

2. *Identify and list as columns the states of nature that can occur.* In this problem there are two states of nature: High Demand and Low Demand. So this results in a 2 by 2 size payoff table.

3. *Identify and list in the appropriate row and column the payoffs.* The payoffs are in customer sales, where the $3, $1, $4, and $-2 million in sales values are the payoffs.

4. *Formulate the problem/model as a payoff table.* The payoff table formulation of the complete problem/model is presented in Table 2. (We will return to this problem again when we discuss how to solve DT problems.)

Table 2. DT formulation of the network topology problem.

Alternatives	States of Nature	
	High demand	Low demand
Ring network	3	1
Star network	4	-2

A New IT Service Product Problem

An information systems manager wants to decide which of three new service products (i.e., the product alternatives, A, B, and C) they should introduce. The firm the manager works in operates on a contractual basis and will know the market for these products with certainty. What cannot be determined is the type of economic environment they will have once they have purchased the IT to support the new service product they are planning. Each service product has a given market potential for profitability based on three types of economic market situations (i.e., the states of nature). The economic market situations in this problem are: recession, growth, and depression. Product A's estimated market will generate $2,000 in profit per day in a recession, $3,500 per day in growth period, and might actually cost the firm $1,000 per day in a depression environment. Product B's estimated market will generate $3,000 in a profit per day in a recession, $5,000 per day in a growth period, and might actually cost the firm $1,500 per day in a depression environment. Product C's estimated market will generate $500 in profit per day in a recession, $3,000 per day in growth period, and $2,000 per day in a depression environment. What is the DT problem/model formulation?

Using the four-step procedure we can formulate this problem as follows:

1. *Identify and list as rows the alternatives to choose from.* There are three alternatives: Product A, Product B, and Product C.
2. *Identify and list as columns the states of nature that can occur.* In this problem there are three states of nature: Recession, Growth, and Depression. So this results in a 3 by 3 payoff table.
3. *Identify and list in the appropriate row and column the payoffs.* The payoffs are the in profit of $2,000, $3,500, $-1,000, $3,000, $5,000, $-1,500, $500, $3,000, and $2,000.
4. *Formulate the problem/model as a payoff table.* The payoff table formulation of the complete problem/model is presented in Table 3.

Table 3. Formulation of new IT service product problem.

Alternatives	States of Nature		
	Recession	Growth	Depression
Product A	2,000	3,500	-1,000
Product B	3,000	5,000	-1,500
Product C	500	3,000	2,000

Once a DT is formulated, the payoff table can be used to analyze the payoffs and render a decision. The methodologies that are used to solve a DT problem vary by type of decision environment. In the next few sections of this chapter will examine differing methodologies covering each of the three decision environments.

Decision-Making Under Certainty

There are many different criteria that can be used to aid in making decisions when the decision maker knows with certainty what the payoffs will be in a given state of nature. Two of these criteria are: "maximax" and "maximin".

Maximax criterion

The maximax criterion is a totally optimistic approach to decision-making. The maximax selection is based on the following steps:

> *1. Select the maximum payoff for each alternative, and then...*
> *2. Select the alternative with the maximum payoff of the*
>   *maximum payoffs from Step 1.* Hence, max-i-max!

To illustrate this criterion we will revisit the topology problem. The solution to this problem is presented in Table 4. As we can seen the maximum payoffs for each of the two alternatives are $3 million and $4 million in sales, respectively. Of these, the $4 million payoff is the maximum payoff, so the max of the max is $4 million with the selection of choosing to build a Star Network alternative.

Table 4. Maximax solution for DT network topology problem.

Alternatives	States of Nature		Max payoff for alternatives	Max payoff of the max
	High demand	Low demand		
Ring network	3	1	3	
Star network	4	-2	4	4

Maximin criterion

The maximin criterion is a semi-pessimistic approach that assumes the worst state of nature is going to occur and we should make the best out of it. The maximin selection is based on the following steps:

> *1. Select the minimum payoff for each alternative, and then...*
> *2. Select the alternative with the maximum payoff of the*
>   *minimum payoffs from Step 1.* Hence, max-i-min!

To illustrate this criterion we will again revisit the topology problem. The solution to this problem is presented in Table 5. As we can seen the minimum payoffs for each of the two alternatives are $1 million and $-2 million in sales, respectively. Of these, the $1 million payoff is the maximum payoff, so the max of the min is $1 million with the selection of the Ring Network alternative.

Table 5. Maximin solution for DT network topology problem.

	States of Nature			
			Max payoff	Max payoff of
Alternatives	High demand	Low demand	for alternatives	the max
Ring network	3	1	1	1
Star network	4	-2	-2	

The answers above to the same problem might cause some concern. How can one criterion suggest one alternative and other criterion suggest still another alternative? Indeed, which alternative is the best? It depends on the selection of criteria that a decision maker or manager chooses to use to guide their decision. If a person is an optimist, they will choose a maximax approach to decision-making and if they are more pessimistic, they might choose the maximin approach. In reality, the best decision comes from looking at all possible criteria. Clearly, though, you can use either criterion or both to justify your particular position on a decision and that is perhaps one of the best ways of using these criteria and their related methodologies. It is important in IT investment decision-making that you consider both optimistic and pessimistic realities in any analysis in order to help explain possible outcomes and eventually gain approval for IT investment alternatives.

Decision-Making Under Risk

There are many different criteria that can be used to aid in making decisions when the decision maker knows the problem they face is a "risk" situation where their states of nature are probabilistic. Many IT investment situation demand the consideration of risk in order to determine or measure the value of IT (see Morrell, 2003; Swierczek and

Shrestha, 2003). In such a decision environment both the origin of the probabilities and the criteria used to make a decision are important.

Origin of probabilities

In a risk problem, probabilities are attached to each state of nature. The sum of these probabilities must add to one. So, when we flip a coin, two payoffs are possible (i.e., a head or a tail), the probability of each is 0.50, and adding the sum of all (or both in this case) alternative probabilities equals 1.0.

Where probabilities come from can include "objective" or "subjective" sources. *Objective source probabilities* include experimental observation of historical behavior (e.g., like counting how many heads in ten flips of coin) or using some statistical formula (e.g., like a theoretical probability distribution). When we use objective methods to determine probabilities we assume:

1. The probability of past events or experiments will follow the same pattern in the future,
2. The probabilities are stable in the process that is being observed; and
3. The sample size is adequate to represent the past behavior.

If you are not comfortable in making these assumptions an alternative way of determining probabilities involves the use of *subjective source probabilities*. This involves having experts make their best guess at what a probability should be for the states of nature. Using this approach to probability assessment requires us to assume the experts used are knowledgeable of the behavior for which they are assessing probabilities and that their judgment is reasonably accurate.

Both objective and subjective approaches of probabilities assessment are well represented in management literature. Both are used to aid in DT problem solving in risk environments.

There are many different criteria that can be used to aid in making decisions in a risk environment. Two of these criteria are: Expected Value and Expected Opportunity Loss.

Expected value criterion

The *expected value (EV) criterion* is determined by computing a weighted estimate of payoffs for each alternative. The expected value criterion is based on the following steps:

1. *Attach the probabilities for each state of nature to the payoffs in each row in the payoff table.*
2. *Multiple the probability in decimal form times each payoff and sum by row.* These values are the expected payoffs for each alternative.
3. *Select the alternative with the best payoff. If the problem has profit or sales payoffs, the best payoff would be the largest expected payoff.* If the problem has cost payoffs, the best payoff would be the smallest expected payoff.

To illustrate this criterion we will again revisit the topology problem. This time lets say the probability of a "High Demand" situation is 40 percent and the probability of a "Low Demand" situation is 60 percent. The probabilities attached to the states of nature change this problem into a risk type decision environment. To compute the expected values the probabilities in percentages are changed to decimal values and multiplied times their respective payoff values. In Table 6 the EV's of each alternative are presented in the last column of the payoff table. As can be seen the best payoff (i.e., maximum expected profit) is for the Ring Network alternative at $1.8 million.

Expected opportunity loss criterion

The expected opportunity loss criterion is based on the logic of the avoidance of loss. The decision using this criterion is based on minimizing the expected opportunity loss (i.e., what you stand to lose if the best decision for each state of nature is not selected). The procedure for computing the values on which this criterion is based involve the following steps:

Table 6. Expected value solution for DT network topology problem.

Alternatives	States of Nature		Expected values
	High demand (40%)	Low demand (60%)	
Ring network	3(0.40)+	1(0.60)=	$1.8 million*
Star network	4(0.40)+	-2(0.60)=	$0.4 million

*Best expected sales payoff.

1. *Determine the opportunity loss values in not making the best decision in each state of nature.* This is accomplished by selecting the best payoff under each state of nature and subtracting all the values in that column from that best payoff (including itself). The result of this difference is called opportunity loss as it represents what you stand to lose if the alternative selected is not the best payoff for that state of nature. The opportunity loss values can be structured into an opportunity loss table represented by the same framework as the DT payoff table.
2. *Attach the probabilities to the opportunity loss values and compute expected opportunity loss values for each alternative by summing the products of the probabilities times their respective opportunity loss values by row.*
3. *Select the alternative with the minimum expected opportunity loss value computed in Step 2.*

The steps to this criterion in solving the DT network topology problem are presented in Tables 7a and 7b.

Decision-Making Under Uncertainty

Decision making under uncertainty means that the decision maker has no information at all on which state of nature will occur. They do know what the payoffs will be, the alternatives and what states of nature can exist. There are many different criteria that can be used to aid in making decisions when the decision maker is facing an uncertain decision environment. Five of these criteria are: Laplace, Maximin, Maximax, Hurwicz, and Minimax.

Table 7a. Step 1 of expected opportunity loss solution for DT network topology problem.

Alternatives	States of Nature	
	High demand (40%)	Low demand (60%)
Ring network	3	1
Star network	4	-2

Step 1. Determine the opportunity loss values in not making the best decision in each state of nature. This is accomplished by selecting the best payoff under each state of nature and subtracting all the values in that column from that best payoff.

Alternatives	High demand	Low demand
Best payoff per state of nature	4	1

The opportunity loss values can be structured into an opportunity loss table represented by the same framework as the DT payoff table below:

Alternatives	States of Nature	
	High demand (40%)	Low demand (60%)
Ring network	3-3=**1**	1-1=**0**
Star network	4-4=**0**	-2-(-2)=**3**

So we can see here under the High Demand state of nature that if the alternative Star Network is selected we will have "0" opportunity loss since this is the best possible payoff in this state of nature. Alternatively, if we select Ring Network we have an opportunity loss of $1 million in sales (i.e., $4 - $3 = $1 of loss) since we could have made $4 million instead of just $3 million with that alternative.

Table 7b. Steps 2 and 3 of expected opportunity loss solution for DT network topology problem.

Step 2. Attach the probabilities to the opportunity loss values and compute expected opportunity loss values for each alternative by summing the products of the probabilities times their respective opportunity loss values by row.

	States of	Nature	
Alternatives	High demand (40%)	Low demand (60%)	Expected opportunity loss
Ring network	1(0.40)+	0(0.60)=	$0.4 million
Star network	0(0.40)+	3(0.60)=	$1.8 million

Step 3. Select the minimum expected opportunity loss value computed in Step 2. The minimum expected opportunity loss is with the Ring Network alternative with a value of only $0.4 million.

Laplace criterion

The *Laplace criterion* is based on the *Principle of Insufficient Information*. We assume under this principle that since no information is available on any state of nature, each is equally likely to occur. As such, we can assign an equal probability to each state of nature, and then compute an expected value for each alternative. The Laplace selection is based on the following steps:

1. *Attach an equal probability to each state of nature.* For example, if we have five states of nature, probability of each state of nature is 20%. If we have two states of nature, the probability of each state of nature is 50%.
2. *Compute an expected value for each alternative as if you are using the "expected value" criterion.*

3. *Select the alternative with the best expected value computed in Step 2.*

To illustrate this criterion we will again revisit for DT network topology problem. The solution to this problem is presented in Table 8.

Table 8. Laplace solution for DT network topology problem.

Alternatives	States of Nature	
	High demand (40%)	Low demand (60%)
Ring network	3	1
Star network	4	-2

Step 1. Attach an equal probability to each state of nature. Since we have two states of nature, the probability of each state of nature is 50 percent or 0.50.
Step 2. Compute an expected value for each alternative as if you would using the "expected value" criterion. Expected value computations:

Ring Network: 3(0.50) +   1(0.50) = $2 million
Star Network: 4(0.50) + (- 2)(0.50) = $1 million

Step 3. Select the alternative with the best expected value computed in Step 2. The best alternative is the Ring Network at $2 million in sales.

Maximin criterion

The *maximin criterion* is the same as it was under certainty. The solution is the same as given before.

Maximax criterion

The *maximax criterion* is the same as was under certainty. The solution is the same as given before.

Hurwicz criterion

The *Hurwicz criterion* is a compromised approach between the maximin and maximax approaches. As Cater (1992) has illustrated, this criterion is an excellent means of making IT investment decisions. In using this criterion the decision maker must subjectively weight the degree of optimism they have in the future. The *coefficient of optimism* is used for this weighting. The coefficient of optimism is on a scale from 0 to 1 and is represented by the Greek letter alpha or a. The closer alpha is to 1, the more optimistic the decision maker is about the future. The *coefficient of pessimism* is 1 - a. Both coefficients are used in the computation of the expected payoffs of each alternative. The Hurwicz selection is based on the following steps:

1. *State the value of alpha or a.*
2. *Determine the maximum and minimum payoffs for each alternative.*
3. *Multiply the coefficient of optimism (i.e., a) times the maximum payoff, multiply the coefficient of pessimism (i.e., 1 - a) times the minimum payoff, and add these values together to derive the expected value for each alternative.*
4. *Select the alternative with the best expected payoff from Step 3.*

To illustrate this criterion we will again revisit the DT network topology problem. The solution to this problem is presented in Table 9.

Minimax criterion

The *minimax criterion* is similar to expected opportunity loss criterion in that it is based on avoidance of loss. The decision using this criterion is based on minimizing the expected opportunity loss. The procedure for computing the values based on the minimax criterion consists of the following steps:

Table 9. Hurwicz solution to the DT network topology problem.

Steps:
Step 1. State the value of α. Let's let α = 0.7. This means that we are more optimistic (i.e., closer to 1).
Step 2. Determine the maximum and minimum payoffs for each alternative.

Alternatives	States of	Nature	Maximum	Minimum
	High demand	Low demand	payoff	payoff
Ring network	3	1	3	1
Star network	4	-2	4	-2

Step 3. Multiply the coefficient of optimism (α) times the maximum payoff, multiply the coefficient of pessimism (1 - α) times the minimum payoff, and add these values together to derive the expected value for each alternative.

> Ring Network: 3 (0.7) +   1 (1-0.7) = \$2.4 million
> Star Network: 4 (0.7) + (-2)(1-0.7) = \$2.2 million

Step 4. Select the best expected payoff from Step 3. The best payoff is with the Ring Network alternative at \$2.4 million in sales.

1. *Determine the opportunity loss values in not making the best decision in each state of nature.* This is accomplished by selecting the best payoff under each state of nature and subtracting all the values in that column from that best payoff. The opportunity loss values can be structured into an opportunity

loss table represented by the same framework as the DT payoff
table.
2. *Determine the maximum opportunity loss values for each
   alternative.*
3. *Select the alternative with the minimum opportunity loss value
   determined in Step 2.*

The steps to this criterion in solving for DT network topology
problem are presented in Table 10.

Multi-Objective Programming Methodology

As we have seen, DT problems are solved using a variety of criteria and
methodologies. All of these methodologies assume that, once a decision
is made, the alternative recommended by the methodology can be
accepted and implemented. Unfortunately in the real world there are
many limitations on what we can afford in terms of funding, personnel,
technology, and so on. These "constraints" on the resources that we have
can in many IT decision situations prevent the successful
implementation, thus rendering an alternative suggested by DT analysis
useless. Therefore, it managers must consider any relevant potential
resource limitation in the IT investment decision so that an alternative
can be suggested that is doable and therefore the best choice, given the
resource limitations of the organization. Fortunately not all IT
investment decision situations require resource constraints to be
incorporated in the analysis, but for those problems that do, a
methodology called, "multi-objective programming" has been developed.
Steuer (1986) defines *multi-objective programming* (MOP) as a
constraint resource methodology that has multiple objectives. While
there are several methodologies that fit this definition, we will examine
one commonly referred to type of MOP in IT investment analysis called,
"goal programming."

Table 10. Minimax solution of the DT network topology problem.

Alternatives	States of Nature High demand (40%)	Low demand (60%)
Ring network	3	1
Star network	4	-2

Step 1. *Determine the opportunity loss values in not making the best decision in each state of nature.* This is accomplished by selecting the best payoff under each state of nature and subtracting all the values in that column from that best payoff.

Alternatives	High demand	Low demand
Best payoff per state of nature	4	1

The opportunity loss values can be structured into an opportunity loss table represented by the same framework as the DT payoff table.

Alternatives	States of Nature High demand (40%)	Low demand (60%)
Ring network	3-3=**1**	1-1=**0**
Star network	4-4=**0**	-2-(-2)=**3**

Step 2. Determine the maximum opportunity loss values for each alternative.

Alternatives	States of Nature High demand	Low demand	Maximum payoff
Ring network	1	0	1
Star network	0	3	3

Step 3. *Select the minimum opportunity loss value determined in Step 2.* The minimum of the maximum opportunity loss values is with the Ring Network alternative with a payoff of $1 million in sales.

What is Goal Programming?

Goal Programming (GP) is a deterministic (i.e., with known constant parameters like DT), multi-variable (i.e., has more than one unknown or decision variable like DT, but unlike DT can solve for more than one alternative and allows proportional choices), constrained (i.e., unlike DT has mathematical expressions that constrain the values of the decision variables), and multi-criteria satisfying methodology (i.e., seeks to solve for a solution that best fits or satisfies the desired set of multi-criteria in a problem situation). Originally developed by Charnes and Cooper (1961), there was well over a thousand GP articles published by the mid-1990's (Schniederjans, 1995). Unique to GP models, are their ability to prioritize or rank the importance of constraints in the solution process. GP is a very diverse methodology, permitting application to just about every possible problem situation that fits the assumptions the model requires (we will discuss the assumptions of the GP model in a later section of this chapter). GP can and has been used to model problems in all the functional areas of business (i.e., accounting, economics, finance, management, and marketing) and in all types of operations (i.e., industry, government, agriculture, etc.). It has been recently identified as an ideal IT investment methodology in strategic decisions (Sarkis and Sundarraj, 2000; Wen, *et al.*, 1998), or for the selection of entire MIS (Schniederjans and Hamaker, 2003). GP models have also been suggested for inclusion in a variety of IT investment evaluation frameworks (Sylla and Wen, 2002).

Goal programming problem/model elements

All GP problem/model formulations consist of three elements: *goal constraints*, an *objective function*, and the *non-negativity* or given requirements. A generalized model (i.e., a model without actual values, only symbols) of the three components is presented in Table 11. The generalized model in Table 11 is used here to foreshadow the formulation of GP models discussed in this chapter and provides a kind of "big picture" of what every GP model formulation looks like. Our purpose in this chapter it to acquaint decision makers with formulation skills of GP modeling, while the solution effort will be left to computer software.

Table 11. A generalized statement of a GP model.

Minimize: $Z = d_1^+ + d_2^+ + \ldots + d_m^+ + d_1^- + d_2^- + \ldots + d_m^-$

subject to:
$$a_{11} x_1 + a_{12} x_2 + \ldots + a_{1n} x_n - d_1^+ + d_1^- = b_1$$
$$a_{21} x_1 + a_{22} x_2 + \ldots + a_{2n} x_n - d_2^+ + d_2^- = b_2$$
$$\vdots$$
$$a_{m1} x_1 + a_{m2} x_2 + \ldots + a_{mn} x_n - d_m^+ + d_m^- = b_m$$

and $x_j, d_i^-, d_i^+ \geq 0$

where:

Z = *an unknown value.*

d_i^+ = *positive deviation variable for i = 1, 2, ..., m, that permits deviation above a desired goal.*

d_i^- = *negative deviation variable for i = 1, 2, ..., m, that permit deviation below a desired goal.*

x_j = *decision variables for j=1,2,.., n; which are the unknowns that we seek to determine. These variables can be just zero or one (used in "yes" or "no" IT investments), integer where multiple units of IT are chosen), or any real number (i.e., fractions permitted for resources like dollar values).*

b_i = *a right-hand-side value for i = 1, 2, ..., m; and m are the number of goal constraints in the model each having a right-hand-side resource value representing a target or goal for that resource.*

a_{ij} = *technology coefficients for i = 1, 2, ..., m and*

j = *1,2,...,n; which represents the per unit usage of the related ith right-hand-side resource value by the related jth decision variable.*

Goal Constraints

Lets take an IT example to illustrate this formulation effort for goal constraints. Suppose we must decide between two systems, A or B. Let's say that system A will cost us $5,000 per month to lease and system B will cost $11,000 per month to lease. Lets further say, that we set our target total monthly lease payment at \$10,000. If we set the two systems to be represented by the zero-one decision variables x1 (for system A) and x2 (for system B), we can express this cost relationship as a goal constraint:

$$5,000x_1 + 11,000x_2 + d_1^- - d_1^+ = 10,000$$

Note that all the GP constraints are equalities. This is always the case because of the addition of two new variables, called *deviation variables* (i.e., d_i^+ and d_i^-), where d_i^+ is the positive deviation variable for $i = 1, 2, ...,$ m, and d_i^- is the negative deviation variable for $i = 1, 2, ..., m$. These variables allow deviation from the right-hand-side value in either direction. The positive deviation variable allows positive deviation from the right-hand-side value and the negative deviation variable allows negative deviation the right-hand-side value. The positive deviation variable will always have a negative sign in front of it and the negative deviation variable will always have a positive sign. This is so because in the goal constraint they are being used algebraically to cancel out the deviation from the right-hand-side values (i.e., goal targets). This is the same as saying that we can have an amount greater than b_i (i.e., having positive deviation from b_i) and having an amount less than b_i (i.e., having negative deviation from b_i). But you cannot have deviation in both directions at the same time (i.e., $d_i^- \times d_i^+ = 0$). Let's illustrate this with simple example. Suppose we have the following goal constraint:

$$x_1 + d_1^- - d_1^+ = 10$$

If $x_1 = 6$, then it must be true that $d_1^- = 4$ and $d_1^+ = 0$, so they sum to 10. Also, if $x_1 = 13$, then it must be true that $d_1^- = 0$ and $d_1^+ = 3$, so they sum to 10. Finally, if $x_1 = 10$, then it must be true that $d_1^- = 0$ and $d_1^+ = 0$, so they sum to 10. In other words, at least one deviation variable will always equal zero, so $d_i^- \times d_i^+ = 0$.

The right-hand-side b values in goal constraints represent goal targets. They also have an added control that allows us to target their usage to achieve our own, multiple objectives for the model and in any IT investment decision. This is way we call right-hand-side b values "targets" or "goals" in a GP model. To fully understand the goal constraints, we need to understand how the objective function can manipulate the goals.

Objective Function's Use for Seeking Goals

As stated in the generalized model in Table 11 the objective function is always going to be an equality and a minimization function. It seeks to minimize absolute deviation from all the goals stated in the GP model. This means that all deviation variables in the objective function will have a positive sign. Since it always seeks to minimize whatever deviation variables are placed into it we can use the selection of the deviation variables to specifically target what we want to accomplish with each goal constraint. We have three choices in seeking right-hand-side targets by choosing one of the three alternative options: (1) to put both deviation variables in the objective function, (2) to put only the positive deviation variable in the objective function, or (3) to only put the negative deviation variable in the objective function. Which of these options we choose will help us to achieve our desired goals. The three choices and their resulting goals are stated in Table 12.

Table 12. Selecting deviation variables for the GP objective function.

Alternative choices	Deviation variable selected to go into the objective function	Goal we seek
1	d_i^-, d_i^+	b_i exactly
2	d_i^-	$\geq b_i$
3	d_i^+	$\leq b_i$

A PC Selection Problem

Let's illustrate the objective function with a simple problem of trying to decide how many units of personal computers (PC's) and PC printers a manager should purchase to service a new PC network. So, the decision variables will be integer (i.e., you have to purchase whole units of PC's and printers):

x_1 = number of units of PC's to purchase
x_2 = number of units of PC printers to purchase

In this problem we recognize that we do have some resource limitations. We observe that there are space limitations. The amount of space a PC or PC printer takes up in a physical facility is same. The office where this IT will be used has a maximum allowable utilization space for efficient usage for both PC's and PC printers of only 50 units (i.e., you can not have more than a total of 50 units of either PC or PC printers or a combination of both in the space allowed). To model this goal constraint we again return to the generalized form of the goal constraint:

$a_{11} x_1 + a_{12} x_2 + d_i^- - d_i^+ = b_i$ (Generalized form)

So, the resulting applied goal constraint for our model given the parameters above is:

$x_1 + x_2 + d_1^- - d_1^+ = 50$ (Physical space)

There is an implied coefficient for a_{ij} of 1 in the constraint above, since the space tradeoff of a PC and a PC printer is 1-for-1. Now suppose the company needs to have at least 10 PC's and 5 PC printers in order to meet minimum demand requires of customers. These requirements have to be modeled with separate constraints as follows:

$x_1 + d_2^- - d_2^+ = 10$ (Minimum PC's)
$x_2 + d_3^- - d_3^+ = 5$ (Minimum PC printers)

Next we are given a budget of $10,000 for the IT project. Let's say the desire (or goal) is to use about $10,000 (i.e., no more, no less than $10,000) though it might turn out that we will use more or less than the $10,000 if other goals at higher priorities necessities additional purchases. Let's say it cost $237 for each PC, and $112 for each PC printer. The goal constraint for this budget limitation is:

$237 \ x_1 + 112 \ x_2 + d_4^- - d_4^+ = 10000$ (Budget)

Having structured the goal constraints with both deviation variables, we next need to determine which of the deviation variables from each constraint should be placed in the objective function. To achieve the desired goals stated for each constraint requires the correct placement of either one or both deviation variables from each constraint in the objective function. As previously stated, all deviation variables in the objective function must have positive signs. Why? Because we want to minimize absolute deviation and not have the positive deviation cancelled out by the negative deviation. What you place in the objective function is the type of deviation we want to minimize or avoid. The objective function to model the PC selection problem includes:

1. Only the d_1^+ variable from the first goal constraint should be included in the objective function because we want to avoid any positive deviation from the right-hand-side value of 50. By putting this variable in the objective function we will hopefully minimize it down to zero, thus allowing what we want, which is <50 units of PC and PC printers in a space that allows only 50 units to operate efficiently.

2. Only the d_2^- and the d_3^- variables should be included in the objective function from the second and third goal constraints in order to achieve the goal of having at least 10 PC's and at least 5 PC printers. By putting these two variables in the objective function we will hopefully minimize them down to zero, thus allowing what we want, which is > 10 PC's and > 5 PC printers.

3. Both the d_4^- and d_4^+ variables should be included in the objective function from the fourth goal constraint because we seek to achieve the $10,000 budget value exactly. By putting both variables in the objective function we will hopefully minimize them both down to zero, thus achieving what we want, not going over or under our budget of $10,000.

Finally, it should be mentioned that GP's modeling flexibility permits decision makers to set arbitrary goals or targets. For example, if we wanted to minimize the values of the decision variables in the constraint below, we could set a goal of "0" and put the positive deviation variable in the objective function:

$$x_1 + x_2 + d_1^- - d_1^+ = 0$$

Alternatively, if we want to maximize the values of the decision variables in the constraint below, we can arbitrarily set a very large value (i.e., 999,999) as a goal and put the negative deviation variable in the objective function:

$$x_1 + x_2 + d_1^- - d_1^+ = 999999$$

Now that we have decided which deviation variable to put in the objective function, we next have to decide which type of objective function we should use to solve the problem.

Types of Objective Functions

There are several different types of objective functions that can be used in GP models. We will look at three of the most common. They are differentiated by the types of options each provide the decision maker. These three model objective functions can be generally expressed as in Table 13.

In Table 13 two new parameters are added to the GP models. The P_K or *priority factors* are a ranking system. These symbols do not have a numerical value in and of themselves. They simply serve to notate the ranking of the deviation variables in order of their importance or in order of the solution process, which seeks to minimize each in their respective goal constraint. The higher the rank, where $P_1 > P_2 > P_3 >>> P_K$, defines the ordering of algorithmic procedure to process them. So, all deviation variables placed at P_1 will be reduced to zero before consideration of those deviation variables at P_2, and so on. This is one of the reasons why GP is referred to as a *preemptive solution process* that seeks a solution with the best deviation minimized values or best multi-criteria satisfying solutions, consistent with the ranking of goal.

Table 13. The different types of GP model objective functions.

If a problem has no mention of rankings or priorities, then it is characterized as a simple non-priority, non-weighted model summing up the $i = 1, 2, .., I$ deviation variables in the model.

$$\text{Minimize:} \quad Z = \sum_{i=1}^{I} (d_i^- + d_i^+)$$

If a problem has rankings or priorities, but does not mention weightings within those priorities, then it is characterized as having $k = 1, 2, ..., K$ priorities but not a weighted model.

$$\text{Minimize:} \quad Z = \sum_{i=1}^{I} P_k (d_i^- + d_i^+) \quad (\text{for } k = 1, 2, ..., K)$$

If a problem has rankings and priorities, as well as weightings within one or more priorities, then it is characterized as having both P_k priorities and w_{kl} weights within priorities.

$$\text{Minimize:} \quad Z = \sum_{i=1}^{I} w_{kl} P_k (d_i^- + d_i^+) \quad (\text{for } k = 1, 2, ..., K; \text{ for } l = 1, 2, ..., L)$$

The other new parameter introduced in Table 13 is the w_{kl} or *differential weighting* (also called *mathematical weights*). These weights are limited to be applied within a single priority level and can be any positive real number. While you can have more than one weighting system in a given model, the weighting will always be related to deviation variables within a single priority level. The purpose of these weights is to rank deviation variables within a single priority (i.e., a rank within a rank). The larger the w_{kl} weight, the higher the ranking (i.e., a weight of 2 is greater than a weight of 1).

Now let's return to the three types of objective functions in Table 13 and let's also change the PC selection problem to use each objective function to deal with a different situation.

Suppose you feel that each of the four constraints are equally important (i.e., each goal is equally important). This situation has no priorities and no weights. Given the previous statements on which deviation variables should be included in the objective function, the specific objective function for this model is:

Minimize: $Z = d_1^+ + d_2^- + d_3^- + d_4^- + d_4^+$

Note that in the objective function all deviation variables have a positive sign because we are seeking to minimize total absolute deviation, both positive and negative. Realize that these five deviation variables have no priorities or special weights. This means that the GP solution procedure will treat each of these deviation variables equally and not process their minimization in any special order.

Now let's say that we feel that our goals should be prioritized so that the first goal of space utilization is considered fully before considering the second goal of having a minimum number of PC's and PC printers. Finally, the budget goal should be considered last. That is, we want to preempt the goals so space is considered first, minimum number of units is considered second, and the budget is considered third. This preemtiveness is achieved with P_K preemptive priorities. We do this by placing the deviation variable for space at the first priority (P_1), the deviation variables for the number of units at the second priority (P_2), and the deviation variables for the budget at the third priority (P_3). This situation with priorities but no weights can be expressed as:

Minimize: $Z = P_1 (d_1^+) + P_2 (d_2^- + d_3^-) + P_3 (d_4^- + d_4^+)$

Note, it is important to remember that these priorities are absolute. Deviation variables at higher priorities are minimized as close to zero as possible before the solution procedure even considers the next lower prioritized deviation variable. Moreover, lower level deviation will never be reduced at the expense of higher-level prioritized deviation. In other words, higher-level goals will not be compromised to reduce deviation at lower level goals.

Now let's say that we feel that our minimum PC and PC printer goals should be weighted to reflect the desire that we are more interested in assuring the PC goal than when compared to the PC printer goal. Let's also say that our personal preference here, and this can be very subjective or objective criteria, is that it is twice as important that we fully achieve the PC goal of having at least 10 PC's as we are in achieving the PC printers goal of having a minimum of 5 printers. This weighting actually serves to rank-within-a-rank or to place the PC goal ahead of the less expensive PC printer goal within the second priority. We accomplish this by placing the mathematical weight (i.e., 2 to 1) on the respective deviation variables at the second priority (P_2). This objective function can then be expressed as:

Minimize: $Z = P_1 (d_1^+) + 2P_2 (d_2^-) + 1P_2 (d_3^-) + P_3 (d_4^- + d_4^+)$

For this model, the mathematical weights could have also been the actual price difference between a PC and a PC printer (i.e., \$237 and \$112, respectively). In that case the objective function would be:

Minimize: $Z = P_1 (d_1^+) + 237P_2 (d_2^-) + 112P_2 (d_3^-) + P_3 (d_4^- + d_4^+)$

Since the relative size of the weight does not matter, either of the above two objective functions would have resulted in the exact same solution. That is, when the GP solution procedure is guided by this objective function it first goes to the P_1 deviation variable (i.e., d_1^+). Then it goes to all the deviation variables at P_2 and looks for weights, and the larger the weighted deviation variable (d_2^- having a larger weight) is minimized first. Then it will move to the next largest weighted deviation variable at P_2 (d_3^- having a weight) to minimize second. The solution procedure would then proceed to the next priority and minimize the deviation variables at that priority.

Which of these types of GP objective function models you use depends on the existence of priorities and/or weights in the problem situation. Not all problems require mathematical weights or priorities.

Non-Negativity and Given Requirements

To complete the GP model formulation it is necessary to include the third model element of *non-negativity* and given requirements. The decision variables and the deviation variables in GP models are required to be zero or some positive value. As a formal part of the correct way of formulating a model we must add the additional statement on GP model formulations that can be generally expressed as:

and x_j, d_i^-, $d_i^+ \geq 0$

This expression tells users that this model permits its decision and deviation variables to be zero or any positive real number value, including fractional values. What if you want to determine whole units for the decision variable values (like the PC selection problem)? That requires the solution to generate only integer values for the decision variables. Those *non-negativity* and given requirements will look like:

and $x_j \geq 0$ and all integer; d_i^-, $d_i^+ \geq 0$

For those unique problems where we want the decision variables to be either zero or one, the *non-negativity* and given requirements will look like:

and $x_j = 0$ or 1; d_i^-, $d_i^+ \geq 0$

Goal Programming Problem/Model Formulation

GP problem/model formulation procedure

The hardest part of formulating any problem is always the first step. This stepwise procedure is a strategy for handling any kind of GP model. Big or small, it handles them all:

1. *Define the decision variables*: Two things to remember in defining the decision variables are: (1) be precise and make clear what your decision variables are determining, and (2) state any "time horizon" the problem is requiring.
2. *Formulate the goal constraints*: For some GP models the goal constraints are actually in the listing of goals or priorities in the problem. In this step focus on what is possible (i.e., deviation in

both directions), not your desired goals as we will build those in later.

3. *Determine the preemptive priorities, if need be*: A problem has to have a very clear listing or numbering of goals for you to assume a ranked priority is included. These may be stated as, "management has the following goals in order of their importance." Some IT problems will just have a listing of goals numbered in an order: 1, 2, ..., to some K number of goals. Remember not all GP problems have priorities.

4. *Determine the differential or mathematical weights, if need be*: Differential or mathematical weights are only present if you have priorities. No priorities, no weights! Also, while these weights can be objective in nature, they tend to be very subjective. You may observe statements like, "weight these goals by their cost or profit coefficients," or maybe, "this goal is twice as important as another goal." In this later case, the weights would be 2 to 1. These weights are always placed in front of the priorities in the objective function but are in reality mathematically attached to the deviation variables.

5. *Formulate the objective function*: This is one of the hardest steps as you are trying to model "desires." The goal constraints state what is possible (in Step 2), but here you want to make clear what your desires and goals are in the goal constraint. Sometimes the goals are given as a listing of objectives and will state their desire to achieve over utilization (referring to d_i^+), under utilization (referring to d_i^-), under achievement (referring to d_i^-), or over achievement (referring to d_i^+). If these goals are what they want, then they should put the opposite deviation variable in the objective function. Why the opposite variable? Because whatever you put in the objective function is what you want to minimize or avoid. Note, if the four terms above have minimize in front of them, you put that deviation variable in the objective function (not the opposite). For example, if we want to "avoid over utilization", we would put the d_i^+ deviation variable in the objective function and if we want "over utilization", we would put the d_i^- deviation variable in the objective function.

6. *State the Non-Negativity and Given Requirements*: This is simply the formal listing of the decision and deviation variables in the model as a statement of non-negativity and given requirements.

GP problem/model formulation problems

Let's practice the formulation procedure on a couple of GP IT problems.

LAN Leasing Problem

An MIS manager is in charge of leasing three local area networks (LAN's) to service a firm's customers. The manager's job is to set up a weekly leasing plan that will seek to accomplish the firm's objectives of providing a full 40 hours of service per week, satisfying their customers' demand who use the computer services from the leased LAN's, and minimizing the firm's costs of leasing the LAN's. The firm's customers each have access to all three LAN's so anyone could provide all the leasing requirements if desired. In this case, the manager feels its better to spread the lease assignments over one or more of the LAN partners to do the best job and with minimum costs. In researching the transaction processing capabilities of the LAN's, the manager finds that each LAN can handle a different number of customers per hour and that the cost of the system varies as stated in Table 14.

The firm has a total leasing cost budget of $12,000 for each week's operation. Also, the maximum number of customers expected in any week is 1,500. The manager must plan to offer a total of 40 hours of service in any week and those hours are to be leased from one or more of the three LAN's. The manager states the following as desired goals of this plan, and they are ranked in order of their importance in the following order:

Table 14. Data for LAN leasing problem.

LAN's	Hourly utilization Maximum number of customers LAN technology can serve	rates IT costs($)
LAN System A	50	350
LAN System B	40	300
LAN System C	25	200

1. There must be exactly 40 hours of service provided each week. That is, they do not need more or less than 40 hours of service from the LAN's.
2. The firm wants to minimize under utilization of the 1,500 customers' demand. In other words, they want to try and achieve at least 1,500 customers served by providing the necessary IT capacity.
3. The firm wants to minimize the over utilization of their $12,000 budget of IT lease costs. In other words, they don't want to go over budget.

Let's use the six-step GP problem/model formulation procedure to structure the GP model for this LAN leasing problem.

1. *Define the decision variables*: We will assume that we are talking about whole hours or an integer solution here. Given that, the decision variables can be defined as:

x_1=number of hours to lease LAN 1 system per week
x_2=number of hours to lease LAN 2 system per week
x_3=number of hours to lease LAN 3 system per week

2. *Formulate the goal constraints*: The formulation of the goal constraints for this model are related to the two limitations in Table 14 and the 40 hours required in the week. The basic form of the goal constraints follows closely to the stated three goals above:

$x_1 + x_2 + x_3 + d_1^- - d_1^+ = 40$ (P_1: Hours in a week)

The goal constraint above states that the sum of the hours allocated to all three LAN's has a goal of a total of 40 hours. Similarly the customers served and the budget goal constraints are formulated by using the data in Table 14.

$50x_1 + 40x_2 + 25x_3 + d_2^- - d_2^+ = 1500$ (P_2: Customers served in a week)
$350x_1 + 300x_2 + 200x_3 + d_3^- - d_3^+ = 12000$ (P_3: Weekly budget for leasing)

3. *Determine the preemptive priorities, if need be:* There are three prioritized goals given in this problem. The first P_1 goal is the 40 hours per week goal, the second P_2 goal is the customer demand service goal, and the third P_3 goal is the budget goal.

4. *Determine the differential or mathematical weights, if need be:* No mathematical or differential weights were stated in this problem, so this problem has no weights.

5. *Formulate the objective function:* We have to re-read the goals again to formulate the objective function. Specifically, the goal statements will tell us which deviation variable or variables should be placed in the objective function. Refer to Table 12 to help you determine which deviation variable should go into the objective function. Below we will look at each goal separately:

> 1. "There must be exactly 40 hours of service provided each week." Here, we don't want to over utilize or under utilize the 40 hour goal target. So we don't want to have either positive or negative deviation from the goal. By placing both the positive and negative deviation variables at P1, we seek to minimize them both. This portion of the GP model's objective function will look like:
>
> Minimize: $Z = P_1 (d_1^- + d_1^+)$
>
> Since this goal is at the highest priority, there will be no conflict in reducing both deviation variables down to zero, resulting in a solution where 40 hours of service will be made available to the firm's customers.
>
> 2. "The firm wants to minimize under utilization of the 1,600 customers demand." To minimize or avoid under utilization, you include the negative deviation variable in the objective function at the P2 priority level. By avoiding negative deviation we leave the opportunity open in the constraint for the positive deviation variable to become as large are possible. Adding this

portion of the GP model's objective function to the last, it will look like:

Minimize: $Z = P_1 (d_1^- + d_1^+) + P_2 (d_2^-)$

3. "The firm wants to minimize the over utilization of their $12,000 budget of IT lease costs." To minimize or avoid over utilization of a goal, we include the positive deviation variable in the objective function at the P3 priority level. By avoiding positive deviation we leave the opportunity open in the constraint for the negative deviation variable to become as large are possible. This remaining portion of the GP model's objective function will look like:

Minimize: $Z = P_1 (d_1^- + d_1^+) + P_2 (d_2^-) + P_3 (d_3^+)$

6. *State the Non-Negativity and Given Requirements*: Since decision variables are integer hour values, they need to be designated as integers, whereas the deviation variable does not require such a designation. The non-negativity and given requirement then become:

and $x_1, x_2, x_3 \geq 0$ and integer; $d_1^-, d_1^+, d_2^-, d_2^+, d_3^-, d_3^+ \geq 0$

The complete formulation of this GP problem/model is presented in Table 15.

Table 15. GP model formulation of the LAN leasing problem.

Minimize: $Z = P_1 (d_1^- + d_1^+) + P_2 (d_2^-) + P_3 (d_3^+)$

subject to:
$$x_1 + x_2 + x_3 + d_1^- - d_1^+ = 40 \text{ (Hours in a week)}$$
$$50x_1 + 40x_2 + 25x_3 + d_2^- - d_2^+ = 1500 \text{ (Customers served in a week)}$$
$$350x_1 + 300x_2 + 200x_3 + d_3^- - d_3^+ = 12000 \text{ (Weekly budget for leasing)}$$

and $x_1, x_2, x_3 \geq 0$ and integer; $d_1^-, d_1^+, d_2^-, d_2^+, d_3^-, d_3^+ \geq 0$

Outsourcing Computer Service Problem

The ABC Data Storage firm outsources its operating systems (OS) software programming and its computer application (CA) software programming needs to XYZ Human Resources firm that specializes in providing hourly services to clients. Outsourcing planning is performed on a yearly basis with an annual budget just for outsourcing programming. The ABC firm has a yearly budget for OS and CA programming set at approximately $52,000 with the possibility of more or less being spent, if needs can be justified. In the upcoming year the XYZ firm plans on charging $180 per hour for OS programming and $150 per hour for CA programming. The ABC firm will need at least 200 hours of OS programming services because they plan on introducing a new update of their operating system software during the upcoming year. The CA software involves little more than maintenance during the next year and the amount of its allocation can be viewed as highly flexible. The firm easily estimates that 130 hours or less for CA programming services will be adequate for their needs for the up coming year. Also, the ABC firm has a labor contract with existing software programmers of the firm that limits outsourcing to a total of 300 hours per year. The ABC firm's management has set the following goals in order of their importance.

1. Avoid under utilization of OS programming.
2. Avoid over utilization of CA programming.
3. Avoid over achievement and under achievement of the existing software limitation of 300 hours.
4. Avoid over achievement and under achievement of the budget. Weight the possibility of going over as twice as important as going under.

Let's use the six-step GP problem/model formulation procedure again to model this outsourcing computer service problem.

1. *Define the decision variables*: This problem has a time horizon of the upcoming year and is focused on allocating outsourcing hours. So the decision variables are:

x_1=number of hours of OS software programming to outsource to XYZ during the upcoming year
x_2=number of hours of CA software programming to outsource to XYZ during the upcoming year

2. *Formulate the goal constraints*: The formulation of the goal constraints for this model is closely related to the four goals listed, in order, in the problem. These are fairly straight forward as can be seen below:

$x_1 + d_1^- - d_1^+ = 200$ (P_1: OS hours available)
$x_2 + d_2^- - d_2^+ = 130$ (P_2: CA hours available)
$x_1 + x_2 + d_3^- - d_3^+ = 300$ (P_3: Maximum hours permitted under contract)
$180x_1 + 150x_2 + d_4^- - d_4^+ = 52000$ (P_4: Budget)

3. *Determine the preemptive priorities, if need be*: This problem has a very clear listing or numbering of goals that can be used as ranked priorities. In this problem the goals were listed 1, 2, 3, and 4, therefore, should be ordered by priorities P_1, P_2, P_3, and P_4.

4. *Determine the differential or mathematical weights, if need be*: Differential weights, w_{kl}, are included in this problem. The mathematical weights used to differentiate the under achievement and over achievement are 2 to 1 for the positive and negative deviation variables, respectively.

5. *Formulate the objective function:* We have to re-read the goals again to formulate the objective function. Specifically, the goal statements will tell us which deviation variable or variables should be placed in the objective function. Refer again to Table 12 to help you determine the appropriate deviation variable to go into the objective function. Below are the goals listed by their priority number:

> 1. At the first priority P1, we want to "Avoid under utilization of OS programming." So we don't want to go under the 200 hours of OS time available. The under utilization deviation variable is

the negative deviation variable. If we want to avoid under utilization, then we would put the negative deviation variable in the objective function since this variable will be minimized. This portion of the GP model's objective function looks like:

Minimize: $Z = P_1(d_1^-)$

2. At the second priority P_2, we want to we want to "Avoid over utilization of CA programming." So we don't want to go above the 130 hours of CA time available. The over utilization deviation variable is the positive deviation variable. If we want to avoid over utilization, then we would put the positive deviation variable in the objective function to minimize it. This adds the next portion of the GP model's objective function, which looks like:

Minimize: $Z = P_1(d_1^-) + P_2(d_2^+)$

3. At the third priority P_3, we want to "Avoid over achievement or under achievement of the existing software limitation of 300 hours." So we don't want to go above or below the 300 hours permitted under contract with the firm's existing programmers. To avoid over or under achievement you place both positive and negative deviation variables in the objective function at the third priority as shown below:

Minimize: $Z = P_1(d_1^-) + P_2(d_2^+) + P_3(d_3^- + d_3^+)$

4. At the fourth priority P_4, we want to "Avoid over achievement and under achievement of the budget." So we don't want to go above or below the $52,000 budget. To avoid over or under achievement you place both positive and negative deviation variables in the objective function. This goal also has mathematical weighting that we want to include, "Weight the possibility of going over as twice as important as going under." Going under is represented by the negative deviation variable, so its weight, $w_{41} = 1$, and going over is represented by the positive deviation variable, so its weight, $w_{42} = 2$. The resulting final component that completes the objective function is:

Minimize: $Z = P_1(d_1^-) + P_2(d_2^+) + P_3(d_3^- + d_3^+) + 1P_4(d_4^-)$
$+ 2P_4(d_4^+)$

6. *State the Non-Negativity and Given Requirements*: In this problem we did not state that hours had to be whole or integer, so they can be fractional and no integer requirement is needed as given below:

and x_1, x_2, d_1^-, d_1^+, d_2^-, d_2^+, d_3^-, d_3^+, d_4^-, $d_4^+ \geq 0$

The complete formulation of the GP model for outsourcing service problem is presented in Table 16.

Table 16. GP model formulation of the outsourcing service problem.

Minimize: $Z = P_1(d_1^-) + P_2(d_2^+) + P_3(d_3^- + d_3^+) + 1P_4(d_4^-) + 2P_4(d_4^+)$

subject to: $x_1 + d_1^- - d_1^+ = 200$ (P$_1$: OS hours available)
$x_2 + d_2^- - d_2^+ = 130$ (P$_2$: CA hours available)
$x_1 + x_2 + d_3^- - d_3^+ = 300$ (P$_3$: Maximum hours permitted under contract)
$180x_1 + 150x_2 + d_4^- - d_4^+ = 52000$ (P$_4$: Budget)

and x_1, x_2, d_1^-, d_1^+, d_2^-, d_2^+, d_3^-, d_3^+, d_4^-, $d_4^+ \geq 0$

Computer-based solutions for goal programming

The solution procedure for GP is a *modified simplex method* (i.e., finite mathematics based on matrix algebra) from a computer program by Lee (1972). Because of the combined ordinal priorities and cardinal computations, simply Excel© solutions are not practical. There are many commercial software applications that exist to run GP models, such as *LINDO* (Lindo Corporation, 2003) and AB:QM (Lee, 1996). We will be letting the computer generate our solution using the *AB:QM* software in this chapter, since its printout is so similar to most commercial packages. Lets look at the solutions to both problems we formulated in the last couple of sections of this chapter.

Let's first look at the solution for the LAN leasing problem from Table 15. The focus here is on the informational value of the solution in the context of the original problem. The AB:QM computer printout of this solution is presented in Figure 1.

Program: Goal Programming
Problem Title : LAN Leasing Problem

***** Input Data *****
Min Z =   1P1d+1   + 1P1d-1   + 1P2d-2   +1P3d+3
Subject to
C1   1X1 + 1X2 + 1X3 + d-1 - d+1 = 40
C2   50X1 + 40X2 + 25X3 + d-2 - d+2 = 1500
C3   350X1 + 300X2 + 200X3 + d-3 - d+3 = 12000
***** Program Output *****
Analysis of deviations

Constraint	RHS Value	d+	d-
C1	40.000	0.000	0.000
C2	1500.000	0.000	0.000
C3	12000.000	0.000	700.000

Analysis of decision variables

Variable	Solution Value
X1	0.000
X2	33.000
X3	7.000

Analysis of the objective function

Priority	Nonachievement
P1	0.000
P2	0.000
P3	0.000

***** End of Output *****

Figure 1. Computer solution for LAN leasing problem.

As we can see, the printout can be initially divided into an ***Input Data*** section and a ***Program Output*** section. The Input Data section restates the input data for the problem. Note that each of the three constraints are labeled C1, C2, and C3, respectively. These labels will be used later.

The Program Output section has three tables. The *analysis of deviations* table provides the optimized values for all the deviation variables. Each of the C1, C2, and C3 constraints are listed with their right-hand-side values. The last two columns list the deviation variable values with columns for both positive and negative deviation variables. As we can see in Figure 1 all three of the positive deviation variables are zero and the first two negative deviation variables are zero. The only variable that is positive is $d_3^- = 700$. So we can interpret this to mean that we had a negative deviation from the $12,000 budget limitation. The negative deviation of 700 is $700 from the total budget of $12,000, or we will only use $11,300 of the $12,000 total budget in the resulting solution.

The second table in the GP printout is called the *analysis of decision variables*, and it provides the optimized decision variable values. In this problem in Figure 1 we will lease 33 hours of the LAN system B (i.e., $x_2 = 33$) and 7 hours of LAN system C (i.e., $x_3 = 7$). This is our solution to the LAN leasing problem.

The third table in the GP printout is called the, *analysis of the objective function*, and represents information on how well we achieved our goals. The term *nonachievement* simply means the deviation that could not be reduced from the targeted goals in the model prevented a full achievement of that goal. In this table, each priority is identified and a related nonachievement value is provided. What this nonachievement value represents is the sum of all deviation for all deviation variables at that priority level. Its interpretation is quite easy to understand. For all goals with zero nonachievement, those goals can be considered as being "fully achieved." For goals with some positive deviation, we would interpret them as "not being fully achieved." The $700 of negative deviation in the analysis of deviation table did not constitute deviation from a goal since we did not include the negative deviation in the objective function. In other words, we didn't care if we went under the

$12,000 budget and we did. In the LAN leasing problem we have fully achieved the all three goals as we have stated them in the objective function.

Let's now look at the solution to the outsourcing service problem from Table 16, which is presented in Figure 2.

Program: Goal Programming
Problem Title : Outsourcing Sevice Problem

***** Input Data *****
Min Z = 1P1d-1 + 1P2d+2 + 1P3d-3 + 1P3d+3 + 1P4d-4 + 2P4d+4
Subject to
C1   1X1 + d-1 - d+1 = 200
C2   1X2 + d-2 - d+2 = 130
C3   1X1 + 1X2 + d-3 - d+3 = 300
C4   180X1 + 150X2 + d-4 - d+4 = 52000
***** Program Output *****
Analysis of deviations

Constraint	RHS Value	d+	d-
C1	200.000	33.000	0.000
C2	130.000	0.000	63.000
C3	300.000	0.000	0.000
C4	52000.000	0.000	10.000

Analysis of decision variables

Variable	Solution Value
X1	233.000
X2	67.000

Analysis of the objective function

Priority	Nonachievement
P1	0.000
P2	0.000
P3	0.000
P4	10.000

***** End of Output *****

Figure 2. Computer solution for LAN leasing problem.

The program output is again organized into three tables. The analysis of deviations table provides the values for all the deviation variables. Each of the C1, C2, C3 and C4 constraints are listed with their right-hand-side values and the last two columns list the deviation variable values for both positive and negative deviation variables. As we can see in Figure 2 all but three of the deviation variables are zero. The value of $d_1{}^+{=}33$ represents the positive deviation of 33 hours of OS time from the 200 hours set as a goal. The value of $d_2{}^-{=}63$ represents the negative deviation of 63 hours of CA time from the 130 hours set as a goal. The negative deviation of 10 is \$10 from the total budget of \$52,000, or we will only use \$51,990 of the \$52,000 total budget. In the analysis of decision variables table in the printout we can see the optimized decision variable values. In this problem in Figure 2 we will end up outsourcing 233 hours of the OS time (i.e., $x_1{=}233$) and only 67 hours of CA time (i.e., $x_2{=}67$). In the analysis of the objective function table we can review the information on how well we achieved our goals. Since we only have a value of 10 at the P_4 fourth priority, it means have fully achieved our goals that P_1, P_2 and P_3, and have only partially achieved our goal at P_4.

One of the interesting things about this solution is the deviation at P_4. Remember that we weighted the $d_4{}^-$ with a "1" and $d_4{}^+$ with a "2", making $d_4{}^+$ a more desirable target for minimization. So, why did the GP model minimize $d_4{}^-$ fully and not fully minimize $d_4{}^+$? The answer is simple: higher level priorities over rode the importance of the weighting at the fourth priority. This ability to handle conflict in limited resources is one of the most appealing features of GP and is one of the reasons why it is currently considered an ideal IT investment decision-making methodology.

Goal Programming Complications and Model Assumptions

GP complications

As Romero (1991) points out, there are a number of complications that can prevent a solution method in a GP model from generating a desired criteria fitting solution or even formulating a problem correctly. By being aware of these complications and what causes them GP users can more easily overcome them. Some of these complications (but not limited to these) are: unbounded solutions, infeasible solutions, dominance in goals, and incommeasurability.

Unbounded Solutions: An unbounded solution is not a solution. It means that you have incorrectly formulated the problem so that one or more of the decision variables in the model goes off to positive infinity. The resolution of this complication is to reformulate the problem correctly.

Infeasible Solutions: It is not possible to obtain an infeasible solution unless you leave both of the deviation variables out of your model. And this means that you have incorrectly formulated the problem in such a way that no solution set could be found that would satisfy all of the constraints in the model. The solution is again to reformulate it correctly.

Dominance in Goals: Selecting goal targets can be very subjective. If, for example, we wanted to minimize a set of decision variables by placing a "0" for a goal target at the highest priority, it would dominate the solution, forcing all the decision variables to zero. Anytime a dominating constraint is set a high priority, it will dominate all the goals under it. Such constraints can also cause unbounded and infeasible solution complications. How can we avoid this problem? Always place any potential dominating constraint at the last priority in any model. If you don't have priorities in you model then create two and place the dominating constraint at the second priority.

Incommeasurability of Goals: Have you heard of the expression that you should not mix apples and oranges except in a fruit cocktail? Well in GP you should avoid putting multiple constraints with different units of measure (i.e., \$ with tons of gold, labor hours with parking spaces, etc.) at the same priority level. Why? In the GP solution procedure the model will seek to minimize total deviation for all deviation variables in a non-weighted priority with equal measure. That is, if we have the two goal constraints below:

$$x_1 + d_1^- - d_1^+ = 1,000,000 \text{ (Dollars)}$$
$$x_2 + d_2^- - d_2^+ = 20 \text{ (Tons of gold)}$$

and if their deviation variables are all placed in the objective function at the same priority as:

Minimize: $Z = P_1 (d_1^- + d_1^+ + d_2^- + d_2^+)$

The solution procedure would focus on minimizing deviation for the "Dollars" goal constraint first, and then minimize the deviation in the "Tons of gold" constraint second. Realistically, tons of gold are much more important on dollar-to-ton basis, yet in the GP model it will treat them as equals in value as they appear only as a unit of deviation. So, what can we do about this unfair weighting of goals? A simple solution is to separate each goal constraint into an individual priority. That permits the decision maker to control the preemptive prioritization process. If it is important for two goals to stay at a single priority, another approach is to convert the unit of measure into the same unit. So, we could convert the tons of gold into dollars for both constraints. This avoids making an unfair comparison between dissimilar units of measure. Hence, the term, *incommeasurability*.

GP model assumptions

There are at least seven basic assumptions that must all be met in order for us to use GP to model any decision-making situation. These assumptions are again:

1. *Linearity*: All constraints and the objective function must be linear functions.
2. *Additivity*: All of the constraints and the objective function must, for any value of the decision variables, add up exactly as modeled. That is, you cannot have any synergistic impact (i.e., where 2+2 equals more than 4).
3. *Divisibility*: In the GP models presented in this chapter the non-negativity and given requirements allow the decision variable values to be real numbers or have fractional values.
4. *Finiteness*: This requirement simply means that the values of the decision variables must be finite.
5. *Certainty and a Static Time Period*: All of the a_{ij}, b_i, and (when applicable) w_{kl} parameters of a GP model must be known with certainty and valid over the time horizon used in the model.
6. *Ordination of goals is absolute*: What this means is that once the preemptive priorities are established and viewed as absolute,

each must be as fully achieved before consideration of others at lower priorities (i.e., $P_1 > P_2 >>>>>> P_K$).

7. *Positive and Negative Deviation can exist for each goal*: When we incorporate both positive and negative deviation variables we assume that in a real world problem deviation is possible in either direction.

Summary

In this chapter we have examined two types of IT investment decision-making methodologies: decision theory methods and goal programming models. Our discussion of decision theory included its use under differing decision environments and with a variety of different criteria. The computation procedures for the various decision criteria were also presented. Our discussion of goal programming was focused on model/problem formulation and solution interpretation.

Both DT and GP have unique strengths and weaknesses (i.e., their assumptions and limitations) in solving multi-criteria IT investment problems. Other methodologies exist that can incorporate multi-criteria information in differing ways to achieve an entirely different type of IT investment analysis that is focused more on comparisons. In Part IV of this book, we start our examination of several additional methodologies that explore unique comparative analyses to provide useful IT investment information on which a decision can be made.

Review Terms

Additivity

Alternatives

Analysis of decision variables table

Analysis of deviations table

Analysis of the objective function table

Certainty

Certainty and a static time period

Coefficient of optimism

Coefficient of pessimism

Decision theory (DT)

Maximax criterion

Maximin criterion

Minimax criterion

Modified simplex method

Multi-objective programming (MOP)

Nonachievement

Non-negativity requirements

Negative deviation variable

Objective function

Objective source probabilities

Decision variables
Deterministic
Deviation variables
Differential weighting
Divisibility
Dominance in goals
Expected opportunity loss
Expected opportunity loss criterion
Expected value (EV) criterion
Finiteness
Goal constraints
Goal Programming (GP)
Hurwicz criterion
Incommeasurability of goals
Infeasible Solutions
Laplace criterion
Linearity
Local area networks (LAN's)
Mathematical weights

Over achievement
Over utilization
Payoff table
Positive deviation variable
Preemptive solution process
Principle of insufficient information
Priority factors
Pure choice problem
Right-hand-side value
Ring network
Risk
Star network
States of nature
Subjective source probabilities
Technology coefficients
Unbounded solutions
Uncertainty
Under achievement
Under utilization

Discussion Questions

1. Where would we use DT methodologies instead of GP problems/models?
2. What determines which DT criteria is used in a particular DT decision environment?
3. Why do we have differing DT criteria for solving problems?
4. What do we do when differing DT criteria suggest differing solutions?
5. Since both expected value criterion analysis in risk decision environments and the Hurwicz criterion in uncertain decision environments use probabilities, why can't they be used in both decision environments?
6. Where would we use GT problem/models instead of DT methodologies?
7. What are "differential weights" used for in GP problems and models?
8. Why is GP considered a "preemptive" solution method?
9. How many priorities are necessary in a GP problem/model? Do we need priorities in GP problems/models?

10. Why do GP problems/models have to meet all of the modeling assumptions stated in this chapter?

Concept Questions

1. What are the three decision environments possible in DT problems?
2. What are the criteria used in making decisions in a DT environment of "certainty"?
3. What are the criteria used in making decisions in a DT environment of "risk"?
4. What are the criteria used in making decisions in a DT environment of "uncertainty"?
5. What is the logic behind the Hurwicz criterion?
6. What are the three basic components of a GP model?
7. Where do we get the "mathematical weighting" for some GP problems/models?
8. Why don't we put all the deviation variables of a GP model in its objective function?
9. What does the "incommeasurability of goals" limitation mean in GP problems/models?
10. What type of information is provided in an "analysis of deviations" table of a AB:QM printout?

Problems

1. A company has just reduced its size through a reengineering process. They now find they have $5 million to reinvest in one of the three types of resources: new personnel, new technology, or new processes. The firm works in an environment that permits no assurance of what the economic environment will be, nor has it information on what the environment will most likely become. The projected profit for an investment in the personnel resource can be either $1.6 million per year if a "prosperous market" exists, or only $0.4 million if a "depressed market" exists. The projected profit for an investment in the technology resource can be either $2.3 million per year if a "prosperous market" exists, or only $1.1 million if a "depressed market" exists. The projected

profit for an investment in the process resource can be either $0.7 million per year if a "prosperous market" exists, or $4.0 million if a "depressed market" exists. Formulate this as a DT problem/model.

2. Is Problem 1 above a certainty problem or uncertainty problem?

3. If we use a Maximax criteria in Problem 1, what is the optimal choice?

4. If we use the Maximin criteria in Problem 1, what is the optimal choice?

5. To meet price competitive a company must decide which one of three different IT projects they want to outsource to four different outsourcing firms. Firm A will charge $34,000 for the first project, $15,000 for the second, and $23,000 for the third. Firm B will charge $37,000 for the first project, $18,000 for the second, and $20,000 for the third. Firm C will charge $39,000 for the first project, $12,000 for the second, and $18,000 for the third. Firm D will charge a flat fee of $20,000 for each project. Formulate this problem as a DT problem/model.

6. The XYZ firm must allocate hours of outsourcing time between two firms A and B. The XYZ firm has the following goals in order of their importance:
   a. To allocate exactly 40 hours between the two firms.
   b. To serve exactly 50,000 customers per week.
   c. Because of differing IT advantages, firm A can process only 420 customers per hour and firm B can process 600 customers per hour.
   d. To allocate no more than the week's budget of $100,000 to cover all outsourcing costs.
   Formulate this problem as a GP model.

7. Given the GP model formulation in Problem 6 and the Program Output portion of the AB:QM printout of its solution below, answer the following questions:
   a. How many hours of outsourcing should be allocated to both firm A and B?
   b. What goals are fully achieved? What goals are not fully achieved?

Analysis of deviations				Analysis of dec. var.		Analysis of the Obj. function	
Constraint	RHS Value	d+	d-	Var.	Sol. Value	Priority	Non achieve-ment
C1	40	0	0	X1	0	P1	0
C2	50000	0	6000	X2	40	P2	26000
C3	100000	0	92000			P3	0

8. A small company sells three products: x-1, x-2, and x-3. The expected monthly demand for the three products has been estimated at 2,000, 2,500 and 3, 400, respectively. Each unit of each product requires the input of several units of three types of materials (i.e., A, B, and C) as stated in the table below:

Units of Material Used			
Product	A	B	C
x-1	10	12	18
x-2	14	11	12
x-3	16	13	17

The company can afford 24,000 units of material A, 30,000 units of B and 32,000 units of C for production each month. These levels should be viewed as production usage targets. More materials can be acquired if need be. The profit when the products are manufactured and sold for one unit of x-1 is $56, one unit of x-2 is $58, and one unit of x-3 is $63. Using the company's goals in order of their ranked importance, formulate as GP model:

1. Produce at least the expected monthly demand as estimated.
2. Underachieve the material production usage targets.
3. Maximize profit.

9. Given the GP model formulation in Problem 8 and the Program Output portion of the AB:QM printout of its solution below, answer the following questions:

******Program Output*****

Analysis of deviations				Analysis of dec var	Analysis of the Obj. function		
Constraint	RHS Value	d+	d-	Var.	Sol. Value	Priority	Non achievement
C1	200	0	0	X1	0	P1	0
C2	100	0	100	X2	200	P2	0
C3	250	0	50	X3	0	P3	50
C4	400	0	400			P4	0

a. How many hours of each labor should be used?
b. What goals are fully achieved? What goals are not fully achieved?
c. What does the d_3^- deviation value represent? Explain.
d. Why didn't we use more of the "really unskilled hours"? Explain.

10. An MIS manager must decide to use one of four systems: A, B, C or D. The system capabilities and costs are given in the table below:

	Transaction	processing	capabilities	and costs
System	Customers served per hour	Maintenance costs per hour ($)	Capital costs* ($)	Useful life in years
A	40	0.90	1,500,000	3
B	50	1.10	1,750,000	3.5
C	75	2.00	2,500,000	5
D	85	2.25	3,000,000	10

*Estimated over the life of the machine

The MIS manager has stated four goals below, in order of their importance:

a. Seek to achieve a minimum of 40 customers served using the IT purchased.
b. Seek to achieve no more than $2 per hour maintenance costs.
c. Seek to achieve no more than a $2 million capital investment in the IT.
d. Seek to achieve at least a 5-year useful life in the IT purchased.
Formulate this problem as a GP model.

References

Charnes, A. and Cooper, W.W., *Management Models and Industrial Applications of Linear* Programming, New York, NY: John Wiley & Sons, 1961.

Lee, S.M., AB:QM Software, Boston, MA: Allyn and Bacon, 1996.

Lee, S.M., *Goal Programming for Decision Analysis*, Philadelphia, PA: Auerbach Publishers, 1972.

LINDO, Lindo software systems, Chicago, IL, 2003, (www.lindo.com).

Meredith, J., Shafer, S. and Turban, E., *Quantitative Business Modeling*, Mason, OH: South-Western, 2002.

Morrell, E., "United by Technology," *Risk Management*, Vol. 50, No. 7, 2003, pp. 46-49.

Moore, J.H. and Weatherford, L.R., *Decision Modeling with Microsoft Excel*, Upper Saddle River, NJ: Prentice Hall, 2001.

Romero, C., *Handbook of Critical Issues in Goal Programming*, Oxford, UK: Pergamon Press, 1991.

Schniederjans, M.J. and Hamaker, J., "A New Strategic Information Technology Investment Model," *Management Decision*, Vol. 41, No. 1, 2003, pp. 8-17.

Schniederjans, M.J., *Goal Programming: Methodology and Applications*, Boston, MA: Kluwer Academic Publishers, 1995.

Savage, S.L., *Decision-Making with Insight*, Belmont, CA: Brooks/Cole-Thomson Learning, 2003.

Swierczek, F.W. and Shrestha, P.K., "Information Technology and Productivity: A Comparison of Japanese and Asia-Pacific Banks," *Journal of High Technology Management Research*, Vol. 14 No. 2, 2003, pp. 269-289.

Sylla, C. and Wen, H.J., "A Conceptual Framework for Evaluation of IT Investments," *International Journal of Technology Management*, Vol. 24, Nos. 2/3, 2002, pp. 236-261.

Wen, H.J., Yen, D. and Lin, B., "Methods for Measuring Information Technology Investment Payoff," *Human Systems Management*, Vol. 17, No. 2, 1998, pp. 145-155.

Part IV

Other Information Technology Investment Methods

Chapter 10

Benchmarking Techniques and Game Theory

Learning Objectives

After completing this chapter, you should be able to:

- Explain what "benchmarking" is and how it can be implemented for information technology planning.
- Explain what "gap analysis" is and how it can be used to identify and graphically display performance issues in information technology management.
- Describe "game theory" as a means of selecting information technology investment alternatives.
- Explain the difference between "pure strategy" and "mixed strategy" decision situations.
- Explain the consequences of information technology alternative investment selections when a "game theory" problem has a "saddle point" solution.

Introduction

In the previous part of this book we have presented a variety of financial and non-financial methodologies useful in *information technology* (IT) investment decision-making. In this first chapter of "Part IV. Other Information Technology Investment Methods" we look at a collection of procedures, guidelines and methodologies that have elements of both financial and non-financial methodologies. In this chapter we examine the comparative process of "benchmarking", which incorporates quantitative performance measure methodology but is itself a non-quantitative process for motivating improvements in IT and the organization as whole. We will also explore the use of "game theory" in IT investment decision-making. This methodology, while highly quantitative, is actually focused on behavior and can incorporate financial or non-financial information as a basis for decision-making.

What is Benchmarking?

When financial measures like, internal rate of return (IRR) or return on investment (ROI) are used to make a decision on IT, they are usually made in the context of prior IRRs and ROIs values. That is, there is an implicit comparison being made. Indeed, even present value analysis requires an interest rate on which a final comparison of time-valued measurements can be made. Virtually all of the methodologies previously presented require some kind of comparison on which to render a final decision.

Benchmarking is an explicit comparative analysis involving the selection of a "best performance" standard for products, services, or practices that represents what the best any individual, group, or organization can do (Laudon and Laudon, 2004, p. 387; Schniederjans and Cao, 2002, p. 196). Once selected, the "benchmark" is used as a target or goal to be achieved. Firms wanting to achieve "world-class" performance in IT acquisition use benchmarking.

There are many ways that benchmarking can be applied to IT investment decision-making to support MIS programs. Some examples of IT benchmarking are presented in Table 1.

For example, Wheat (2003) reported on a Cap Gemini Ernst & Young study that established investment benchmarks for heath managed organizations (HMOs). In this study, per member of cost averages and statistics were provided. Breaking down the areas of IT investment into categories like claims, sales and marketing, network management, information systems, etc, they then provided mean and upper and lower range statistics (e.g., "Claims" in 2002 had a mean of \$1.14 per member, a \$1.51 lower 25th percentile, and a \$1.29 for the upper 75th percentile). The breakdowns of the dollars invested in IT by category allow HMO IT managers to see how well they are doing relative to competitors in their own individual IT investments. Spending above or below the mean values, and particularly the boundary percentile per member in any HMO might call for an investigation for reasons to justify those outcomes.

Table 1. Benchmarking measures in IT.

IT areas for benchmarking	Measures
Provider of IT services	-Average initial time to respond to requests for IT service -Average elapsed time to completion of IT projects -Average customer (external and internal) satisfaction score using IT -Average number of defects per project, transaction, etc. -Quality measures or ratings by operations staff, auditors, customers, suppliers, vendors, etc. -Response time to requests for services -On-time delivery rate
Efficiency in the use of IT resources	-Late charges paid to vendors or suppliers -Average number of days required in training personnel -Cost per project, per transaction, per department -Total budget over run costs -Percentage of lost business due to IT failure
Profitable operations	-Return on IT assets -Profit/revenue per employee
Human resource management	-Employee satisfaction scores in use of IT -Employee turnover rates -Employee absence rates

A benchmarking procedure

A suggested procedure for benchmarking usually includes the following steps (Schniederjans and Cao, 2002, pp. 203-204; Shah and Singh, 2001; Spendolini, 1992; Zee, 2002, pp. 142-164):

1. *Establish a benchmarking team to oversee the implementation of the benchmarking process.* Benchmarking requires oversight to insure success. The team should be guided by supervisors and staff from the same areas where process change will take place. This might mean that members inside and outside of the firm should be a part of the team. The teams should also include technical specialists (i.e., industrial management or systems engineers) who can help to plan the implementation of the processes that will be changed.

2. *Identify the activities or processes requiring improvement.* This may mean defining the scope of the project to be undertaken and areas within the firm or outside that need to be improved. One of the best ways to do this step is by using performance measures for business operations, like those financial and non-financial discussed throughout this book. This can also include any subjective service criteria from customer surveys and from an organization's critical success factors (see Chapter 7 for more information on critical success factors). Since most of these measures are computed routinely for other reasons (i.e., a balanced scorecard method from Chapter 7, or just the usual financial statements of the firm), they do not usually require a great deal of effort to obtain. On the other hand, the area of concern might not have any measure taken on it and thus points out a reporting weakness that should be addressed with new measures.

3. *Identify "best performance"activity or process and measures.* Benchmarking requires the identification of an individual, group, or company whose performance is the "best" in the industry. Some of these firms can be identified by reviewing research reports in the literature (i.e., journal publications, trade magazines, association publications, etc.). One such journal is the *International Journal of IT Standards and Standardization Research.* There are many sources for this information online

(see www.bettermanagement.com or www.industryweek.com). Performance measures should also be collected on these best practices for comparison. Other sources of information on benchmarking include governmental bodies, trade associations, and academic institutions.

4. *Collect data on current operation activities and perform comparative analysis.* The measures collected on current operations can then be compared with those "best performance" benchmarks from an industry leader for differences. The greater the difference defines the ranking of importance (i.e., the greater the difference, the more important that activity or process performance area needs to be improved).

5. *Establish a set of recommended process changes.* Long-term and short-term changes should be defined. Also multiple strategies should be suggested for change implementation. From these alternatives, the "best" of the best should be selected and undertaken.

6. *Follow-up.* To insure the successfulness of this type of program, "visual management" techniques, such as "gap analysis" should be employed. (We will discuss gap analysis in the next section.) Performance measures that helped to identify operations problems, both current and proposed should be posted where related personnel can see them. As progress is made towards the stated benchmark goals, management should communicate the progress and continue to offer suggestions on approaches to improve. Also, updating "best performance" measures should take place periodically as standards change over time.

A benchmarking program is not a one-time process improvement activity, but is meant as a long-term program of what quality managers call a *continuous improvement program.* It is a program of incremental improvement toward meeting, and perhaps beating, performance expectations of customers and benchmarked competitors.

A benchmarking methodology: Gap analysis

To help managers identify and clearly communicate areas where differences between benchmarks and actual performance may exist, a graphical aid called "gap analysis" can be used. *Gap analysis* allows members of an organization to evaluate their performance in achieving benchmarks or any goals. Building on the principle of *visual management* (i.e., the idea that business performance and each individual's contribution to that performance should be openly displayed for all to see) a *gap chart* can be developed that compares an organization's actual performance on the basis of where they are (i.e., the actual status of a technology, system or human resource) and where they desire themselves to be (i.e., a benchmark).

This gap chart can be developed in a number of ways, one of which is to use judgmentally generated measures by giving questions, like those in Table 2 to users of IT services or IT customers. Note, a customer here might be the typical outside customer who purchases an organization's services or it might be internal staff who are customers of IT groups in the organization. The survey, given to customers, will ask them to evaluate various IT service products on select performance measures (e.g., quality, usability, etc.). A rating or scoring method can be used to scale the perceived "actual performance." A rating or score on a 1 (i.e., poor rating) to a 10 (i.e., perfect rating) scale can be used, or any other continuous scale can be used.   This survey will establish the "actual" status of the level of service being provided. By then taking the mean of the individual's scores of each survey, points can be plotted on a gap chart as presented in Figure 1.

The collected benchmarks from the benchmark procedure (Step 3. *"Identify "best performance"activity or process and measure"*) from an industry leader can then be plotted as the organization's target of performance. The difference between the plotted actual and desired points on the chart represents the "gap" between where an organization is and where they want to be. The greater the gap, the greater the necessity for corrective action to be used to bring the performance measure back toward a desired target of performance.

Table 2. Typical customer questions for a gap analysis.

IT performance measure	Survey question
Quality service	How would you rate our service quality?

Poor				Average					Perfect
1	2	3	4	5	6	7	8	9	10

Technology	How would you rate the availability of the IT used?

Poor				Average					Perfect
1	2	3	4	5	6	7	8	9	10

Human resources and staff support	How would you rate our staff service department representative?

Poor				Average					Perfect
1	2	3	4	5	6	7	8	9	10

Communication support	How would you rate the speed of our e-mail communications support?

Poor				Average					Perfect
1	2	3	4	5	6	7	8	9	10

Cost effectiveness	How would you rate the costs of IT support on your budget?

Poor				Average					Perfect
1	2	3	4	5	6	7	8	9	10

As we can see in Figure 1 the gaps between the performance measures of technology, communication, and costs are very minor, indicating that in these areas the organization appears to be providing a level of expected or desired service performance to the customer consistent with the desired benchmark best practices in the industry. The wider gaps between the IT performance measures of quality service and human resources clearly indicate that help is needed in these areas. That help can take the form of either a reengineering program (explained in Chapter 2) undertaken by the organization or by bringing in the necessary consulting expertise to develop programs, training or new procedures to lesson the gap by improving organizational performance.

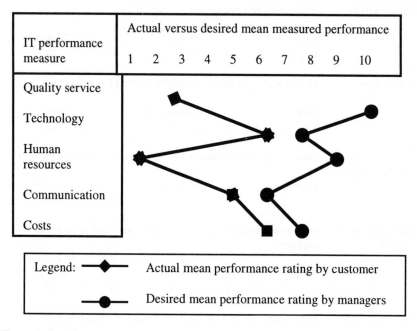

IT performance measure	Actual versus desired mean measured performance									
	1	2	3	4	5	6	7	8	9	10
Quality service										
Technology										
Human resources										
Communication										
Costs										

Legend: ◆ Actual mean performance rating by customer
● Desired mean performance rating by managers

Figure 1. Gap chart.

Like most visual management approaches, gap analysis should be a part of a continuous improvement process that organizations undertake. Periodic changes in perceptions may necessitate an occasional reengineering effort.

What is Game Theory?

Game theory (GT) is an extension of *decision theory* (DT) which was discussed in Chapter 9. Like DT problems, GT problems have multiple *strategies* or *alternatives* to choose from; they have *payoffs*, and can even have *states of nature*. What differentiates GT problems from DT problems? One differentiating factor is that a GT problem exists in a "conflict" type of decision environment. (You will remember DT problems exist in certainty, risk, or uncertainty environments.) In *conflict situations*, there is more than one person making a selection. Like a game of chess; as one player moves, the other player responds to the previous move. This game of moves and counter-moves continues until the game is at an end.

There are many IT situations where understanding and responding to competitive moves or behavior is essential. As such, GT is a valuable IT investment methodology. For example, when a competitor is planning on buying a new IT that might give them a competitive advantage, a competitive firm has to decide if they should likewise acquire the technology to remain competitive. That is a game theory type of problem situation. Butterfield and Pendegraft (2001) have shown how GT can be specifically used for analyzing information system investments in highly volatile competitive environments.

Another factor that differentiates GT from DT problems is that GT allows for "mixed strategy" decisions where proportions of more than one strategy or alternative can be selected (e.g., 75 percent of alternative 1 and 25 percent of alternative 2). You will remember that in DT problems we selected only one strategy or alternative. When only one strategy is selected, it is called a "pure strategy" decision. GT can solve both pure strategy and mixed strategy games.

One final point that is important in using GT is that it is an *optimization method*, generating a solution that is optimal and can not be improved upon. Optimization methods are rare in IT investment decision-making since there is little in the way of perfect benchmarks for comparisons to prove optimality. In GT problems, a compromise is the optimal solution possible if both players are playing a perfect game to win. GT solution methodologies give the exact strategy or combination of strategies to achieve that compromised solution.

Basic structure of game theory problem/model

While game theory problems can have two or more players, we will limit our discussion to what is called a "two-person, zero-sum game." In a two-person, zero-sum game, you have two players and what one player gains, the other loses. The basic structure of the two-person, zero-sum game problem /model is presented in a generalized expression in Table 1. (As we did in Chapter 9, we will again use the combined term "problem/model" to describe a problem formulation that is also used to express its model representation.) This is considered the most common GT type of problem. In fact most business or IT decisions are basically a two-person, zero-sum game. This is so because any one company in an industry, can view themselves as opposing the rest of the industry (i.e., them verses us) and in say the case of market share, what one company gains, the rest of the companies in the industry lose. So in that sense, business decision-making in general is a two-person, zero-sum game.

Table 3. General statement of the two-person, zero-sum game theory problem/model.

	Player B alternatives	or strategies	or states	of nature
Player A strategies	B_1	B_2	...	B_n
A_1	P_{11}	P_{12}	...	P_{1n}
A_2	P_{21}	P_{22}	...	P_{2n}
:	:	:	...	:
A_m	P_{m1}	P_{m2}	...	P_{mn}

In Table 3 the *strategies* that each player can choose can also be called *alternatives* or *states of nature*, like in a DT problem. Player A's strategies (i.e., A_1, A_2, ..., A_m; for A_i, where i=1, 2, ..., m strategies) are always represented by rows in the problem/model formulation, and Player B's strategies (i.e., B_1, B_2, ..., B_n; for B_j, where j=1, 2, ..., n strategies) are represented by columns. The intersection of the rows and columns define the *payoffs* (i.e., P_{ij}; where i=1, 2, ..., m; j=1, 2, ..., n) for any combined two player strategy selection. This table is also called a *game theory payoff table* and represents the problem/model formulation to the GT problem. Similar to a DT problem, the payoffs can be positive, zero or negative. The payoffs can be financial or non-financial such as rating scores or percentages.

In Table 4 an applied GT problem is presented. The game payoffs are always expressed in terms of what Player A "gains" and what Player B "loses". In Table 4, if Player A selects strategy A_1 and Player B selects strategy B_1, the payoff is a gain of 10 to Player A and a loss of 10 to Player B. Likewise, if Player A selects strategy A_2 and Player B selects strategy B_2, the payoff is 0 gain to Player A and a loss of 0 to Player B. If Player A selects strategy A_1 and Player B selects strategy B_2 (i.e., the payoff of -3), the payoff is a loss to Player A of 3 and a gain of 3 to Player B. So a negative value in the payoff table is actually a gain for Player B.

Table 4. Applied example of a GT problem/model formulation.

	Player B alternatives	or strategies	or states of nature
Player A strategies	B_1	B_2	B_3
A_1	10	-3	6
A_2	2	0	7
A_3	4	3	0

One easy way of conceptualizing a two-person, zero-sum GT problem is in terms of a business organization negotiating their labor contracts. Suppose management and labor sit down to negotiate a labor contract where labor demands a wage increase of 5 percent per year (i.e., one possible strategy). If labor gains a 5 percent increase, management gives up, pays, or loses the 5 percent. This is a typical two-person, zero-sum game.

Objectives and assumptions

There are two objectives in a GT problem: (1) maximize the payoffs to Player A and, (2) minimize the losses to Player B. GT is an optimization procedure that achieves an optimal compromise between these two cross-purposed objectives. We want to be able to advise either player on which individual strategy to select or proportioned combination of strategies either player should select to achieve their own optimal compromise solution. The optimal solution is called the *value of the game* for the players. When a player selects a single or individual strategy we call it a *pure strategy* solution. If proportions of several strategies are selected we call it a *mixed strategy* solution. The sum of the proportions for either player must add up to one. Just as the pure strategy represents 100 percent of a players' choice, so must the sum of the mixed strategies add up to 100 percent.

One underlining assumption that must hold true to make the GT problem solution work is that both players always make choices in their own best economic interest. That is, Player A always makes choices that will maximize payoffs and Player B always makes choices to minimize payoffs. One unique aspect of a GT solution is its ability to force both players to achieve the optimal compromised solution. Specifically, once the optimal solution is devised and followed by one player, it doesn't matter what the other player chooses to do, both players will receive their respective optimal compromised payoffs. (This equality of payoffs will be illustrated in a later problem.)

We also assume that an opponent can be a real person, a group of all other companies in an industry, or even a state of nature. While it may seem illogical to assume a "state of nature" will behave in its own best interest, if we use a GT approach to model the problem, then we are assuming that an optimal compromise is the best solution we can expect from that state of nature. Considering the unpredictability of IT in almost any state of nature, it may be a very logical assumption that an optimal compromise is our best hope of a solution from a selection of IT alternative strategies.

Game theory problem/model formulation procedure

The GT problem/model formulation procedure consists of the following steps:

1. *Determine Player A's strategies.*
2. *Determine Player B's strategies.*
3. *Construct GT payoff table (row and column headings).*
4. *Enter payoff values to complete GT problem/model formulation.*

The solution methods for the GT problem will be described in the next section. Let's now look at formulating a couple of GT problem/models.

An IT ROI Investment Problem

A company can have a good year, bad year, or a fair year (i.e., three states of nature). An IT manager would like to determine which of three types of online purchasing systems they should invest in based on their return on investment (ROI) possible payoffs. Some online technologies work well in good years or with a high volume of business transactions environment, some actually do better in bad years or with a low volume of transactions, and some do better in fair years or with an average volume of transactions. If the online purchasing Technology A is implemented in a good year it will generate a ROI of $5 in profit per customer order, $4 in a bad year and $6 in a fair year. If the online purchasing Technology B is implemented in a good year it will have an ROI of $2 in profit per customer order, $3 in a bad year, and $7 in a fair year. If the online purchasing Technology C is implemented in a good year it will have a ROI of $4 in profit per customer order, $3 in a bad year, and $0 in a fair year. Formulate this as a GT problem. Using the four-step procedure we can formulate this problem as follows:

1. *Determine Player A's strategies.* In this problem Player A's strategies are investing in the three Technologies (i.e., A, B, or C).
2. *Determine Player B's strategies.* In this problem Player B's strategies are the three states of nature: good, bad, and fair.
3. *Construct GT payoff table (row and column headings).* The GT payoff table headings are in Table 5.

4. *Enter payoff values to complete GT problem/model formulation.* The resulting GT problem/model formulation of the IT ROI investment problem is also presented in Table 5.

Table 5. GT problem/model formulation of IT ROI investment problem.

	Player B states of nature		
Player A strategies	Good(B₁)	Bad(B₂)	Fair(B₃)
Technology A (A₁)	5	4	6
Technology B (A₂)	2	3	7
Technology C (A₃)	4	3	0

Technology Gaming Problem

Two companies, A and B, compete for the information systems services in a small city. The two companies currently split the market for all services, each with 50 percent. A new technology is announced by a manufacturer that will give a decided competitive advantage to the purchaser. Both companies are trying to decide if they should invest in the new technology. If only one company invests in the new technology, it will capture 90 percent of the total market. If both purchase the new technology they will continue to split the market evenly. Structure this game in terms of Company A's market share payoffs. We will call this first formulation, "Formulation 1". In a second formulation, structure the GT problem in terms of what Store A would receive above versus what it is currently receiving (i.e., above 50 percent). We will call this second formulation, "Formulation 2".

Using the four-step procedure we can formulate this problem as follows:

1. *Determine Player A's strategies.* In this problem Player A's or Company A's strategies are the two strategies: "Purchase" or "Not Purchase".
2. *Determine Player B's strategies.* In this problem Player B's or Company B's strategies are the same two strategies as Player A's: "Purchase" or "Not Purchase".

3. *Construct GT payoff table (row and column headings).* The GT payoff table for Formulation 1 can be seen in Table 6. The GT payoff table for Formulation 2 can be seen in Table 7.

4. *Enter payoff values to complete GT problem/model formulation.* For "Formulation 1" the resulting GT problem/model formulation of the IT ROI investment problem is also presented in Table 6. This formulation is in terms of what Company A will actually receive in market share. As can be seen in the GT payoff table, if both purchase or both do not purchase, they will continue to split the market evenly (i.e., Company A will receive a total of 50 percent of the market). If Company A purchases the technology and Company B does not, Company A will receive a total of 90 percent of the market. If Company A does not purchase the technology and Company B does, Company B gains 90 percent of the market while Company A receives a total of only 10 percent.

Table 6. GT problem/model formulation of IT ROI investment problem.

	Company B's strategies	
Company A's strategies	Purchase(B_1)	Not Purchase(B_2)
Purchase (A_1)	50	90
Not Purchase (A_2)	10	50

For "Formulation 2", the actual formulation requires the payoffs to be adjusted to reflect the above 50 percent of the market that Company A is currently receiving. This formulation is easily provided by simply subtracting 50 percent from each of the values in the GT payoff table in Table 6. The resulting GT problem/model formulation is presented in Table 7.

Table 7. Revised GT problem/model formulation of IT ROI investment problem.

	Company B's strategies	
Company A's strategies	Purchase(B_1)	Not Purchase(B_2)
Purchase (A_1)	0	40
Not Purchase (A_2)	-40	0

Solution Procedures for Game Theory Problems

Rational choice method: A saddle point solution

A game theory (GT) problem sometimes has a solution where both players can select a single strategy choice, also called a "pure strategy" choice. We call this type of GT problem solution a *saddle point* solution because it represents a point when reached, both players can not rationally move from the single strategy they have selected. To illustrate the saddle point solution lets solve a GT problem based on a simple "rational choice method". The *rational choice method* is based on both players following the logic of selecting their best choice, based on available choices open to them during each move in the game. The steps to the rational choice method are as follows:

1. *Player A must select the strategy with the single best (i.e., Player A always maximizes) payoff regardless of Player B's payoffs.* Either player can begin with the first move. Let's begin with Player A. (If we had begun with Player B, Player B would select the minimum value.)
2. *Now it is Player B's turn to choose. Given Player A's choice of strategy, Player B would select the minimum payoff in that specific strategy that Player A has chosen.*
3. *Now it is Player A's turn again. Given Player B's choice of strategy, Player A would select the maximum payoff in that specific strategy that Player B has chosen.*
4. *Repeat steps 2 and 3 until the players can no longer move from their last choice. This will result in the value of the game.* The numerical game value called the *value of the game* can be found either at the intersection of the row and column of the "pure strategy" solution by the players, or within the interval of the two differing values of the same row and column in the GT payoff table (indicating a "mixed strategy" solution).

Let's illustrate this procedure by solving the IT ROI investment problem (previously formulated in Table 5). Using the four step solution procedure above we obtain the following:

1. *Player A must select the strategy with the single best (i.e., Player A always maximizes) payoff regardless of Player B's payoffs.* Suppose that we start from Player A's point of view. Which strategy should Player A choose? Since Player A's objective is to maximize payoffs, Player A would start the game by selecting the strategy that would possibly bring with it the largest payoff (i.e., the bracketed payoff of 7 in Table 8). So, to get the payoff of 7 Player A would select strategy A_2.

Table 8. First step in GT IT ROI investment problem solution.

Player A strategies	Player B states of nature		
	Good(B_1)	Bad(B_2)	Fair(B_3)
Technology A (A_1)	5	4	6
[Technology B (A_2)]	2	3	[7]
Technology C (A_3)	4	3	0

2. *Now it is Player B's turn to choose. Given Player A's choice of strategy, Player B would select the minimum payoff in that specific strategy that Player A has chosen.* Player B's objective is to minimize the loss of payoffs. Player B sees that Player A has selected strategy A_2 and responds, like in a game of chess, by selecting strategy B_1. Why B_1 and not B_3? The game began with a move from Player A. In conflict situations, the players respond to each other's moves. Since Player A had selected A_2, the choices for Player B to choose from were based on the payoffs in the A_2 strategy (i.e., payoffs of 2, 3, or 7). The minimized value of these three payoffs is achieved by selecting strategy B_1 with a payoff of 2. Put another way, it is better for Player B to lose 2, than lose 3 with strategy B_2 or lose 7 with strategy B_3 as shown in Table 9.

Table 9. Second step in GT IT ROI investment problem solution.

Player A strategies	Player B states of nature		
	[Good(B$_1$)]	Bad(B$_2$)	Fair(B$_3$)
Technology A (A$_1$)	5	4	6
Technology B (A$_2$)	[2]	3	7
Technology C (A$_3$)	4	3	0

3. *Now it is Player A's turn again. Given Player B's choice of strategy, Player A would select the maximum payoff in that specific strategy that Player B has chosen.* Now Player A sees that Player B has selected B$_1$, and in response selects the choice that will maximize payoffs of the B$_1$ strategy (i.e., payoffs of 5, 2, or 4). The payoff maximizing strategy for Player A is A$_1$ at 5 in Table 10.

Table 10. Third step in GT IT ROI investment problem solution.

Player A strategies	Player B states of nature		
	Good(B$_1$)	Bad(B$_2$)	Fair(B$_3$)
[Technology A (A$_1$)]	[5]	4	6
Technology B (A$_2$)	2	3	7
Technology C (A$_3$)	4	3	0

4. *Repeat steps 2 and 3 until the players can no longer move from their last choice.* This will result in the value of the game. Now Player B sees that Player A has selected A$_1$ and responds with the selection of B$_2$ that will minimize the loss of the A$_1$ strategy (i.e., the B$_2$ strategy will result in a loss to B of only 4, verses the other choices of losing 5 with the selection of strategy B$_1$ are a loss of 6 with the selection of strategy B$_3$) in Table 11.

Table 11. Fourth step in GT IT ROI investment problem solution.

Player A strategies	Player B states of nature		
	Good(B$_1$)	[Bad(B$_2$)]	Fair(B$_3$)
Technology A (A$_1$)	5	[4]	6
Technology B (A$_2$)	2	3	7
Technology C (A$_3$)	4	3	0

Now, Player A sees that Player B has selected the B$_2$ strategy and responds with the same choice as in Step 3 of A$_1$. Why? Of the three choices possible only the strategy of A$_1$, with a payoff of 4, is greater than the other two with a payoff of only 3. So, Player A stays with strategy A$_1$. It would not be in Player A's best interest to move to any other possible strategy. (Remember the basic assumption that GT problems require players to behave in their own best economic interest.) And finally, Player B seeing that Player A stays with the A$_1$ strategy will also stay with the B$_2$ strategy. Why? Because to move to any other strategy will increase loss. Since both players can agree on a single strategy (i.e., Player A selects A$_1$ and Player B selects B$_2$) the result is a "saddle point" solution to the game.

In this, or any other game, the solution includes not only the selection of the strategy but the determination of the "value of the game." The value of the game in a saddle point game is found at the intersection of the two selected strategies (i.e., intersection of the rows and columns selected). The value of this game is 4. Player A gains 4 and Player B loses 4. Notice in the first moves of the game for both players that the values start high for Player A (i.e., at 7) and low for Player B (i.e., at 2). As the game proceeds, the values become compromised to an agreeable value of 4.

Would it have made any difference if we had started with Player B's position? No! As can be seen in Table 12 the moves starting from Player B's position result in the same solution as before. If a game has a saddle point, as in this example, the value of the game is the same for both players. It represents the best possible compromise between the two players seeking to achieve their own personal objectives.

Table 12. Solution if Player B moves first in IT ROI investment problem.

Move	Player	Possible payoffs	Strategy choice	Final choice payoff
1	B	0, 2, 3, 4, 5, 6, 7	B_3	0
2	A	0, 6, 7	A_2	7
3	B	2, 3, 7	B_1	2
4	A	2, 4, 5	A_1	5
5	B	4, 5, 6	B_2	4
6	A	3, 4	A_1	4

Saddle points are not present in every GT problem. For example, "mixed strategy" games (discussed later) can not be solved using the solution method discussed above.

The IT ROI investment problem again presented in Table 13 also illustrates the important concept in GT called, "dominance." *Dominance* in GT problem solutions occurs when one strategy is always preferable to a single other strategy. A dominated strategy is one that a player should never select, as its selection would be a violation of the assumption that a player will always act in their own best interest. Look in Table 13 at the strategy A_1 and A_3. We can say that strategy A_1 dominates A_3 because each payoff for Player A in strategy A_1 is better than each payoff for strategy A_3 (i.e., 5 is better than 4, 4 is better than 3, and 6 is better than 0). A dominated strategy is a redundant strategy since a player will never select it.

The value of dominance is in the fact that dominated strategies can be dropped from a formulation without it impacting the solution. In other words, you can reduce the size of a GT problem by reducing the dominated strategy rows and columns for the formulations. Also, dominance has to be a one-on-one comparison. That is, only one strategy is used to dominate a single other strategy. Dominance can occur for either player in a game but it is not present in every problem.

Table 13. Dominance in the GT IT ROI investment problem strategies.

Player A strategies	Player B states of nature		
	Good(B_1)	Bad(B_2)	Fair(B_3)
Technology A (A_1)	5	4	6
Technology B (A_2)	2	3	7
Technology C (A_3)	4	3	0

The minimax solution method

The *minimax solution method* is a simple step-wise procedure that can generate an optimal solution for a GT problem if it has a saddle point. The minimax solution method can also approximate the solution for mixed strategy problems, but can not determine the exact value of the game. For mixed strategy games it provides a range approximation of the value of the game.

The procedure for the minimax solution method consists of the following steps:

1. *Place the maximum payoff value in each column in a row called "column max."*
2. *Place the minimum payoff value in each row in a column called "row min."*
3. *Select the strategy for Player A that will maximize the payoffs in the row min column.*
4. *Select the strategy for Player B that will minimize the payoffs in the column max row.*
5. *Determine the value of the game.* If the value of the game is the same for both players and falls at the intersection of the selected strategy row and column, the game has a saddle point. If the value of the game (i.e., the row min value and column max value) is different for both players, the true value of the game

will fall between the two payoff values, and the game has a mixed strategy solution.

You can see in this procedure that we are selecting strategies that minimize the maximum payoffs and maximize the minimum payoffs. Hence the name, "minimax solution method".

Let's use the minimax solution method to solve the IT ROI investment problem again as follows:

1. *Place the maximum payoff value in each column in a row called "column max."* The column max values are 5, 4, and 7, since 5 is the maximum value in the B_1 column, 4 is the maximum value in the B_2 column, and 7 is the maximum value in the B_3 column. These values are presented in Table 14.

Table 14. Steps in using minimax solution method for the IT ROI investment problem.

Player A strategies	Player B states of nature			
	Good(B_1)	Bad(B_2)	Fair(B_3)	Row Min
Technology A (A_1)	5	4	6	[4]
Technology B (A_2)	2	3	7	2
Technology C (A_3)	4	3	0	0
Column Max	5	[4]	7	

2. *Place the minimum payoff value in each row in a column called "row min."* The row min values are 4, 2, and 0, respectively, since 4 is the minimum value in the A_1 row, 2 is the minimum in the A_2 row, and 0 is the minimum value in the A_3 row. These values are presented in Table 14.

3. *Select the strategy for Player A that will maximize the payoffs in the row min column.* The maximum payoff of the row min values for Player A is 4. This value is in brackets in Table 14. This means that Player A should select strategy A_1 to maximize payoff gain.

4. *Select the strategy for Player B that will minimize the payoffs in the column max row.* The minimum payoff of the column max values for Player B is also 4. This value is in brackets in Table 14. This means that Player B should select strategy B_2 to minimize payoff loss.

5. *Determine the value of the game.* Since the payoff values of the two strategies selected are the same (i.e., the same intersection value), these values are the value of the game for both players. This game has a saddle point and as such both players have optimally selected pure strategies to achieve it.

A Minimax Solution for a Mixed Strategy Problem

Let's look at another GT problem. This time we examine an IT market share investment GT problem. Suppose we have two companies (i.e., A and B) that are planning IT investments. These company's are the only competitors in the computer services markets in which they compete, so what Company A gains, Company B loses. Company A has an opportunity to take some of the market share away from Company B if they choose to invest money in a collection of new information technologies or lease them. The market share gains to Company A for either of these two strategies are presented in Table 15. Regardless of Company B's response (which can be to ignore the Company A's purchase, lease the same IT system, or outsource the work), they will lose market share because of a current competitive disadvantage. So for example in Table 15, if Company A selects the Purchase IT strategy and Company B selects the Ignore strategy, Company A will receive a market share increase of 4 percent and Company B will lose 4 percent. Let's solve this problem using the minimax solution method:

1. *Place the maximum payoff value in each column in a row called "column max."* The column max values are 5, 7, and 4, respectively. These values are presented in Table 16.

Table 15. Formulation of the IT market share investment GT problem.

	Company B	states	of nature
Company A strategies	Ignore (B₁)	Lease IT (B₂)	Outsource (B₃)
Purchase IT (A₁)	4	7	3
Lease IT (A₂)	5	2	4

Table 16. Steps using minimax solution method for the IT market share investment problem.

	Company B	states	of nature	
Company A strategies	Ignore (B₁)	Lease IT (B₂)	Outsource (B₃)	Row Min
Purchase IT (A₁)	4	7	3	[3]
Lease IT (A₂)	5	2	4	2
Column Max	5	7	[4]	

2. *Place the minimum payoff value in each row in a column called "row min."* The row min values are 3 and 2. These values are presented in Table 16.

3. *Select the strategy for Player A that will maximize the payoffs in the row min column.* The maximum payoff of the row min values for Player A is 3. This value is in brackets in Table 16.

4. *Select the strategy for Player B that will minimize the payoffs in the column max row.* The minimum payoff of the column max values for Player B is This value is in brackets in Table 16.

5. *Determine the value of the game.* Since the intersection values of the two strategies selected are different, the game does not have a saddle point, and the game will require a "mixed strategy" solution. The minimax solution does reveal an interval value in which the true optimal value of the game will eventually be fall. The value of the game in the IT market share investment GT problem has a payoff that falls between 3 and 4 (note these are the row and column values for the selected strategies).

A computer-based solution for a mixed strategy problem

Once it has been determined that the value of the game requires a mixed strategy solution a more complex analytic procedure is necessary to compute the exact proportions of the mixed strategies to be selected. There are several analytic procedures based on matrix algebra that can be used to derive the mixed strategy proportions. Regardless of procedure used, the proportions are best derived by computer because of the computational effort necessary. For those interested in a basic review of some of these mathematical procedures see Hillier and Lieberman (2001, Chapter 14), Render and Stair (2000, Supplement 1), and Weiss (2000, pp. 83-86). For a more extensive discussion of the mathematics see Kelly (2002), Montet *et al.* (2003), and Rasmusen (2000). The Weiss (2000) software application is one of many that are commercially available for solving GT problems.

As we did in Chapter 9 for goal programming, we will again use the AB:QM (Lee, 1996) software to generate a solution for the IT market share investment GT problem. The AB:QM computer-based solution to the GT problem originally presented in Table 15 is presented in Table 17. The AB:QM computer printout of this solution is presented in Figure 1. As we can see, the printout can be initially divided into an ***Input Data*** section and a ***Program Output*** section. The Input Data section restates the input data for the GT problem.

The Program Output section states the solution proportions (referred on the printout as "probability") of each strategy to select. What does the A_1 proportion of 0.333 and A_2 proportion of 0.667 mean? This is the mixture of these two strategies that Player A should select if Player A is to maximize the payoff. Whatever the economic value of these two strategies, Player A should use 33.3% of A_1 and 66.7% of A_2. For example, let's say Player A (which is Company A in the IT problem) has $100,000 to invest in the new technology. To maximize the payoff, Company A should invest $33,333 in strategy A_1 (i.e., Purchasing the IT) and invest the rest, $66,667, in strategy A_2 (i.e., Leasing the IT). The optimal mixed strategy selection for Player B (or Company B) in this game is 0.000 of B_1, 0.167 of B_2, and 0.833 of B_3.

Program: Game Theory
Problem Title : IT market share investment GT problem
***** Input Data *****

A \ B Strategy 1 Strategy 2 Strategy 3

Strategy 1   4   7   3
Strategy 2   5   2   4

***** Program Output *****

Mixed Strategy

For Player A:
Probability of Strategy 1   0.333
Probability of Strategy 2   0.667
For Player B:
Probability of Strategy 1   0.000
Probability of Strategy 2   0.167
Probability of Strategy 3   0.833
Value for this game is 3.67
***** End of Output *****

Figure 2. Computer solution for the IT market share investment GT problem.

Note the occurrence of a proportion of 0 percent for any strategy denotes a "dominated" strategy. A dominated strategy should never be selected by its player. Sometimes dominated strategies in a mixed strategy game are not obvious until the computer solution is generated. In the problem in Figure 2 we can see that strategy B_3 dominates B_1, and is confirmed with the 0 percent in the computer solution.

The value of this game to both players is 3.67 percent, which falls, as expected from the minimax solution, between 3 and 4. Company A gains 3.67 percent market share and Company B loses 3.67 percent.

One of the magical features of game theory problems with mixed strategies is that as long as one player follows their optimal proportions, both players will always receive the exact value of the game. Let's illustrate this magical property of GT. We know that Player B should optimally select 0.000 of B_1, 0.167 of B_2, and 0.833 of B_3. Player B can not rationally ever select B_1, but could select just B_2 or B_3. Suppose Player B just selects B_2 instead of the optimal proportions? The value of the game to Player B is computed below using the proportions Player A selected and the payoffs Player B now selects:

Value of the game = 7(0.333) + 2(0.667) = 3.67

What if Player B instead just selects B3 instead of the optimal proportions? This Player B's game value would be computed as:

Value of the game = 3(0.333) + 4(0.667) = 3.67

In fact, it doesn't make any difference what Player B selects (i.e., a pure strategy or proportions of strategies), the value of the game will be the same as long as Player A uses the optimal proportions.

Summary

This chapter presented benchmarking and game theory methodologies for use in IT planning. Benchmarking, while conceptually a comparative methodology for motivating corrective behavior in a firm, does have quantitative support methodology in the form of gap analysis. Together they can be used to move an organization from a weak to a strong competitive position in IT planning. Game theory was shown as a means of achieving an optimal compromise in conflict investment decision situations with two players. Game theory methods for formulation and solution were also discussed. Some of the solution methods included rational choice, minimax method, and a discussion on computer-based solutions.

Both of the methods presented in this chapter tend to be very non-financial. In the next chapter we examine two methods commonly associated with financial analyzes: "benefit/risk analysis" and "investment portfolio analysis". Both of these methods can utilize financial information but tend to be non-financial in their procedures, making them a type of "other" IT investment methodology characteristic of this Part of this book.

Review Terms

Alternatives	Mixed strategy
Benchmarking	Optimization method
Conflict situations	Payoffs
Continuous improvement program	Pure strategy
Decision theory (DT)	Rational choice method
Dominance	Saddle point
Game theory (GT)	States of nature
Game theory payoff table	Strategies
Gap chart	Two-person, zero-sum game
Information technology (IT)	Value of the game
Minimax solution method	Visual management

Discussion Questions

1. How can benchmarking be used in IT decision analysis?
2. Why do you think we have to establish a benchmark team for a successful application of this methodology?
3. How is "gap analysis" related to "visual management"?
4. How is GT different from DT?
5. What is "dominance" in GT problem solutions?
6. What does a "saddle point" solution really mean?
7. What is the difference between a "pure strategy" and "mixed strategy" solution?
8. If the value of the game is the same for both player's, why are their strategies in a GT solution different?

Concept Questions

1. How can benchmarking use financial information? Explain.
2. What are some examples of benchmarking measures useful in IT? List three.
3. Where should we look for benchmarks? List and explain.
4. How do we prepare a "gap chart"?
5. When formulating a GT problem what types of choices can be made? List three and give an example of each.
6. Why is the "value of the game" equal for both players? That is, why is the value of gain for Player A equal to Player B's loss?

7. Why must the proportions of "mixed strategy" solutions add to one?
8. Which of the methodologies will work to generate a "pure strategy" solution and which will work for a "mixed strategy" solution?

Problems

1. Given the following GT problem, what is the optimal strategy selection for Player A? What is Player B's optimal strategy selection? Use the rational choice method to find the saddle point.

	Player B strategies		
Player A strategies	(B_1)	(B_2)	(B_3)
(A_1)	12	15	0
(A_2)	50	14	12

2. Using the minimax solution method what is the solution to Problem 1? Show your work by listing the choices as shown for example in Table 16 in this chapter.
3. Given the following GT problem, what is the optimal strategy selection for Player A? What is Player B's optimal strategy selection? Use the rational choice method to find the saddle point.

	Player B strategies		
Player A strategies	B_1	B_2	B_3
A_1	-1	20	0
A_2	70	3	2
A_3	112	5	1

4. Using the minimax solution method what is the solution to Problem 3? Show your work by listing the choices as shown for example in Table 16 in this chapter.

5. Given the following GT problem, what is the optimal strategy selection for Player A?

	Player B strategies		
Player A strategies	B_1	B_2	B_3
A_1	0	-40	-110
A_2	50	10	-50
A_3	100	60	10

6. Suppose we want to allocate $10,000 into one or more of three possible IT investment alternatives: A_1, A_2, and A_3. The possible monthly return on investment in these technologies is dependent on the market conditions that may occur. Suppose we have the three market conditions of "above average", "average", and "below average". The percentage return on investment (ROI) for A_1 is estimated to be 6, 2, and 1, for the three market conditions "above average", "average" and "below average", respectively. The percentage ROI for A_2 is estimated to be 2, 3, and 2, for the three market conditions, respectively. The percentage ROI for A_3 is estimated to be 3, 1, and 5, for the three market conditions, respectively. Formulate this as a GT problem/model. Clearly define the GT payoff table headings.

7. Answer the following questions based on the IT investment problem formulated in Problem 6.
   a. Try solving this problem using the minimax solution method. (Hint this problem has a mixed strategy solution so you will be looking for an interval answer.)
   b. (This problem requires software capable of solving GT problems.) What is the mixed strategy solution to this problem? Explain how the $10,000 investment should be allocated over the alternatives?

8. Two banks own Automatic Teller Machines (ATMs). Both ATMs, located across from each other, share a market of several city blocks. Although they both sell essentially the same product, there are two distinct price levels: special service transactions

(high) and economy transactions (low). If both ATMs charge the same price for their services, they split the market evenly. If one is high while the other is low, the low-price ATM will get 70 percent of the business. What is the formulation for this GT problem situation?

9. Answer the following questions based on the IT investment problem formulated in Problem 8.
   a. What is the solution of this game using the "rational choice method"?
   b. What is the solution using the "minimax solution method"?
   c. What, if any, is the difference in strategies?

10. A small town has only two computer technology stores, Store 1 and Store 2. These are the only two stores in the region. The total number of customers is equally divided between the two stores. Assume that a gain of customers by Store 1 is a loss to Store 2, and vice versa. Both stores plan to run annual pre-holiday sales. Sales are advertised through the local newspaper, radio, and television media. With the aid of an advertising firm, Store 1 was able to estimate their daily dollar gains/losses via the differing advertising media. These values are presented in the payoff table below:

	Store 2 strategies		
Store 1 strategies	Newspapers (B₁)	Radio (B₂)	Television (B₃)
Technology A (A₁)	30	40	-80
Technology B (A₁)	0	15	-20
Technology C (A₃)	90	20	50

The values in the table represent $1,000's of dollars in sales per day of the advertising campaign. Assume that Store 1 has $10,000 to allocate for advertising.
   a. What is the GT formulation of this problem?
   b. (This problem requires software capable of solving GT problems.) Find the optimal allocation of Store 1's advertising budget.

References

Butterfield, J. and Pendegraft, N., "Analyzing Information System Investments: A Game-Theoretic Approach," *Information Systems Management*, Summer, 2001, pp. 73-82.

Hillier, F. and Lieberman, G., *Introduction to Operations Research*, 7th ed., Burr Ridge, IL: McGraw-Hill/Irwin, 2001.

Kelly, A., *Decision-Making Using Game Theory*, UK: Cambridge University Press, 2002.

Laudon, K.C. and Laudon, J.P., *Management Information Systems: Managing the Digital Firm*, 8th ed., Upper Saddle River, NJ: Prentice Hall, 2004.

Lee, S.M., *AB:QM Software*, Boston, MA: Allyn and Bacon, 1996.

Montet, C., Dexter, C. and Serra, D., *Game Theory and Economics*, New York, NY: Palgrave/McMillan, 2003.

Rasmusen, E., *Games and Information: An Introduction to Game Theory*, 3rd ed., Malden, MA: Blackwell Publishers, 2000.

Render, B. and Stair, R M., *Quantitative Analysis for Management*, 7th ed., Upper Saddle River, NJ: Prentice Hall, 2000.

Schniederjans, M.J. and Cao, Q., *E-commerce Operations Management*, Singapore: World Scientific, 2002.

Shah, J., and Singh, N., "Benchmarking Internal Supply Chain Performance: Development of a Framework," *The Journal of Supply Chain Management*, Winter 2001, pp. 37-47.

Spendolini, M.J., *The Benchmarking Book*, New York, NY: American Management Association, 1992.

Weiss, H.J., *DS For Windows*, 2nd ed., Upper Saddle River, NJ: Prentice Hall, 2000.

Wheat, H., "Premium Increases Begin to Level Out In Some Areas; Others See No Relief," *Managed Care Outlook*, Vol. 16, No. 23, 2003, pp. 1-8.

Zee, H.V.D., *Measuring the Value of Information Technology*, Hershey, PA: Idea Group Publishing, 2002.

Chapter 11

Investment Portfolio Methodologies

Learning Objectives

After completing this chapter, you should be able to:

- Describe different types of investment portfolio methodologies used in IT investment decision-making.
- Describe Ward's portfolio approach and how it can be used to set priorities in IT investments.
- Describe Peter's portfolio mapping methodology and how it is used to map IT investment strategies.

Introduction

Just as individuals may have portfolios of stocks and bonds, so to do companies have portfolios of technology. In this chapter we examine a set of methodologies that are all related to the idea of establishing a collection or "portfolio" of investments. The idea of portfolio managing of information technology (IT) has spawned a number of differing methodologies. We will examine several of these alternative methods, including Ward's portfolio approach and Peter's mapping of investment methodologies.

What are Investment Portfolio Methodologies?

Investment portfolio methodologies may be defined as IT investment evaluation techniques based on portfolio management. *Portfolio management* involves collecting all IT investments in one place and controlling them as a single set of interrelated activities (McNurlin and Sprague (2004, pp. 70-78). The concept was developed by Harry Markowitz in the late 1950s who later won a Nobel Prize. According to Markowitz organizations may have thousands of IT investment projects underway at any single point in time. These projects may overlap one another and/or be redundant, wasting valuable resources of an organization. Portfolio management provides a means to monitor and manage all IT investments of an organization so that benefits, costs and risks of individual investments can be assessed to determine whether or not they are making a significant contribution to organizational performance.

Portfolio management also allows an overall view of IT investments to evaluate if individual investments correspond with organizational strategy and with one another so that efforts are not duplicated and resources are not wasted.

Portfolio management is a continuous process that views IT investments as assets as opposed to costs of an organization. The IT investment assets must be managed, and monitored continually to make investment and divestiture decisions. Diversification is a portfolio management tool used to decrease or minimize risk.

Investment portfolio methodologies may be best applied in situations where the number of individual IT investments is large and when aligning IT investment with organizational strategy is important. Portfolio management has been applied to IT investment evaluation and management. Two specific methodologies, "Ward's portfolio approach" and "investment mapping methodology" are discussed in the remainder of this chapter.

Ward's Portfolio Approach

Ward's portfolio approach may be defined as an IT investment decision-making technique that views an organization's IT investments as belonging to different categories in a portfolio of IT investments. Each category of IT investments is associated with specific types of benefits and therefore correspondingly appropriate evaluation techniques. Ward's portfolio approach recognizes that subjective evaluation techniques may be best used for a particular category of IT investment, while objective evaluation techniques may be best for a different category of IT investment. This portfolio technique provides a consistent set of rules and procedures with which to evaluate individual IT investments and provides information necessary for setting priorities for a set of IT investments. It is argued by Ward (1990) that an organization must follow a consistent evaluation approach so that it makes consistent decisions with respect to IT investments. Employing a consistent approach allows decision-makers to effectively distinguish between worthwhile and worthless IT investments.

Ward's portfolio approach involves segmenting an organization's IT investments into four categories, identifying possible IT investments in each category and making decisions as to which investments are best. Ward's portfolio approach may be used as an overall IT management technique to evaluate individual IT investments and to prioritize a set of independent IT investments.

Ward (1990) categorizes IT investments by the role they play in the organization and the expected contribution they will make to business performance. Accordingly, the categories of IT investments are as follows: (1) *strategic* (i.e., IT investments critical to the success of the organization), (2) *high potential* (i.e., IT investments that may be important in attaining future success), (3) *factory* (i.e., IT investments the organization currently depends on for success), and (4) *support* (i.e., IT investments valuable but not critical to success).This categorization is intended to reveal the relationship between the IT investment and business success, and to give direction as to how to evaluate and manage IT investments, individually and as a whole. Each category is associated with a specific type of IT investment, intended to fulfill specific objectives and expected to provide specific benefits. The particular category of IT investment dictates the kind of benefits that should materialize the evaluation techniques appropriate to justify an individual

IT investment, and to prioritize a set of IT investments. Let's look at each of the four categories in detail.

1. *Strategic investments* are those that are essential in achieving business objectives and executing strategies. Strategic investments in IT are ones that facilitate change within the organization in an attempt to achieve objectives and attain competitive advantage. As a result, the benefits of strategic investments tend to occur by means of innovating and restructuring internal business processes and external relationships. For this type of IT investment, one will be able to determine the direct costs and possibly some benefits of the investment in quantitative terms; however, in many instances, important intangible benefits and costs cannot be expressed in quantitative terms. These intangibles may be expressed as critical success or failure factors. The importance of these factors, determined by management judgment, to the overall success of the organization will be the basis of deciding whether or not to invest in a particular IT or to prioritize a set of strategic IT investments. The best way to manage strategic IT investments is to incorporate business planning with IT investment planning, so that strategic IT investments are considered in conjunction with business issues and strategies. Typically a steering group with representatives from different organizational levels will evaluate strategic IT investments.

2. *High potential investments* are investments that may or may not be important to business success. These IT investments emerge because of the development of a new business objective or technological advancement. High potential IT investments tend to be associated with more risk than the other types of investments because their benefit are unknown, and thus it is unclear whether or not the benefits will be important to improving organizational performance. One purpose of evaluation is to identify the benefits of the IT and to analyze its expected effect on business performance. This type of investment may be seen as an R & D investment. Consequently, they should be developed and implemented quickly and cheaply to determine whether or not it is an important and/or essential

investment for business success. Often, a *champion* supports and argues for the organization to make a high potential IT investment and must evaluate and present evaluation results to management who, in turn, decides whether or not to invest. If accepted and successful, some high potential IT investments may become strategic investments, "factory investments" or support investments. Other high potential investments may be rejected at the onset, and still others may be scraped after prototype failure. In each case, risk is an important factor to consider in evaluation, as well as the benefits and costs of the investment.

3. *Factory investments* may be defined as those that improve the effectiveness and efficiency of business processes and activities. These are IT investments that the organization depends on for successful execution of operations and support of the design, development, production and delivery of products and services. In general, effectiveness deals with doing the right things. Effectiveness with respect to factory IT investments, is concerned with selecting the right IT to support the right business processes and activities. If the IT supports the right business processes, overall organizational performance should improve. Efficiency has to do with doing the right things right. Once factory IT have been selected to improve the effectiveness of business processes, these IT should also support improving the efficiency with which the business processes are carried out. Again improving the efficiency of business processes and activities should improve overall organizational performance. Many of the benefits of factory IT investments tend to be tangible so that some form of cost/benefit analysis may be conducted to evaluate them. However, in some instances the most influential benefits may be intangible and a critical success factor approach (discussed in Chapter 7) may be used in conjunction with cost/benefit analysis (discussed in Chapter 6). Factory IT investment benefits tend to arise from accomplishing tasks quicker and/or with fewer resources and from linking business activities. Ward (1990) suggests using a strict "feasibility study" approach to evaluating factory IT investments. A *feasibility study* involves analyzing whether or not an

investment is "feasible" given organizational resources and constraints. Three types of feasibility must be addressed:

a. *Technical feasibility*: Do hardware, software and technical resources exist to implement IT?
b. *Economic feasibility*: Do benefits of IT outweigh its costs?
c. *Operational feasibility*: Does proposed IT fit into the existing managerial and organizational context?

Ward (1990) also recommends using a centralized approach in which, standard check sheet items are used to evaluate all factory investments in a consistent manner. The centralized approach to evaluation allows business management to make decisions that are consistent over time with respect to factory IT investments.

4. *Support investments* are IT investments that improve the efficiency of support activities throughout the organization. Support activities are ones that provide the foundation for performing primary business processes and activities of an organization. Support IT investments are valuable, but not critical to business success. The main benefit of a support investment is to improve the efficiency of activities, which often translates into improving productivity. Improvements in productivity tend to be easily measured in quantitative terms, allowing some form of cost/benefit analysis to be the most appropriate evaluation technique. The many different forms of cost/benefit analysis are presented throughout this textbook and include any method that compares quantitative costs with quantitative benefits of an IT investment. Just as with factory investments, some benefits of support investments may be intangible and an appropriate evaluation technique should be used in conjunction with cost/benefit analysis to evaluate support investments. In some instances a centralized decision-making group evaluates support investments, in other instances, a specific department may be charged with the evaluation task.

Evaluating individual investments

Ward's portfolio approach involves several steps that may be conducted to evaluate individual IT investments. The first step is to construct a portfolio of current IT investments. A team within the organization must decide which IT investments belong to which of the four IT investment categories. To accomplish this task, Ward (1990) suggests determining the strategic contribution of each IT investment. The strategic contribution may be measured as the degree to which IT investments meet overall organizational objectives and *critical success factors* (CSFs). (CSFs are discussed in Chapter 7.) The team must first identify overall organizational objectives, the CSFs and then determine to what degree each IT investment contributes to meeting them. Objectives and CSFs are usually identified by conducting personal interviews with several top managers.

The next step is to determine the degree to which each IT investment assists in meeting these objectives and CSFs. One method to do this, as suggested by Ward (1990), is to determine if an IT investment makes a high, medium or low strategic contribution to meeting organizational objectives and CSFs. Strategic IT investments will contribute highly to meeting objectives and CSFs, while high potential IT investments possibly will make a medium to high contribution. Factory IT investment will make a medium to low contribution and support IT investments will make little or a low contribution to meeting objectives and CSFs. Although subjective, this process allows for consistency in categorization by performing the same categorization procedures on every IT investment.

The last step in Ward's portfolio approach is to evaluate an IT investment. As previously discussed, a particular category of IT investment lends itself to be evaluated by particular types of evaluation techniques. Strategic IT investments are characterized as having important intangible benefits in addition to less important, quantifiable costs and benefits. As such, evaluation techniques that consider both tangibles and intangibles may be most appropriate for strategic IT investments. Ward (1990) suggests using CSFs analysis to determine the effects of strategic IT investments on organizational performance. Other methods like mutli-factor scoring methods (Chapter 8), the balanced scorecard (Chapter 7), and multi-objective (Chapter 9), multi-criteria methods may also be useful evaluation tools.   High potential IT investments tend to be risky and have benefits and costs that are unknown to decision makers.   As a result, the evaluation technique selected should incorporate risk into the analysis and be an exploratory technique allowing decision-makers to investigate potential benefits and costs.   Techniques like multi-objective, multi-criteria methods, Delphi method (Chapter 7), benefit/risk analysis, and process quality management may be appropriate techniques.

Factory IT investments are ones intended to improve operational performance of an organization.   The benefits and costs of factory IT investments tend to be tangible so cost/benefit analysis (Chapter 6), net present value analysis (Chapter 4), return on investment or other financial techniques (Chapter 5) may be most appropriate.   In some cases, the most important benefits of a factory investment may be intangible.   In such cases, another method that incorporates intangibles may be more appropriate; Ward (1990) suggests using CSF analysis. Support IT investments usually are intended to improve some aspect of productivity for an organization and improvements in productivity usually may be measured in quantitative terms.   Consequently, just as with factory IT investments, support IT investments may be evaluated with financial techniques, such as, cost/benefit analysis, net present value analysis, payback period or accounting rate of return.   Some important benefits of support IT may also be intangible and these benefits should be considered in the analysis.   It is likely, that they be incorporated in a separate analysis to support the financial evaluation.   It should be recognized that organizations typically have accepted evaluation techniques that are customarily utilized to evaluate particular types of IT investments.   Consideration should be given to customary techniques when selecting an evaluation technique.

Setting priorities

The discussion up until now has focused on describing the unique nature of the different categories of IT investments and on the evaluation of individual IT investments. Ward (1990) has identified procedures specific to each IT category to select an IT investment among a set of independent investments, in other words, to set IT investment priorities. The procedures may be used set priorities to select one IT investment from a set of IT investments and to prioritize IT investments of a particular type. Ward suggests considering: (1) IT benefits, (2) IT costs, and (3) IT risks for each individual investment when appropriate and then setting priorities based on a set of priority rules. The rationale for setting priorities of support IT investments and strategic IT investments is relatively simple. For support IT investments, those with the greatest economic benefit that use the least amount of resources should get highest priority. In other words, those IT investments with the highest ratio of benefits to costs we should be invested in before other investments with lower ratios are considered.

Let's illustrate the procedure above with an example. Suppose an organization must prioritize its support IT investments and select one from a set of three. Table 1 presents that ratio of benefits to costs for each support investment alternative and the corresponding priority for investment.

Table 1. Priorities for a strategic IT investment problem.

	Benefit/Cost ratio	Priority
Support Computer System A	$\frac{1,500,000}{1,000,000} = 1.5$	2nd
Support Computer System B	$\frac{4,000,000}{2,000,000} = 2$	1st
Support Computer System C	$\frac{1,000,000}{800,000} = 1.25$	3rd

According to this analysis, Support Computer System B is the best alternative and should be acquired before investing in the other two. Investment in Support Computer System A should be undertaken second and investment in Support Computer System C should be undertaken third, assuming that no other alternative investment opportunities arise.

Notice that risk has not been considered in this analysis. If deemed appropriate risk may be considered by analyzing it separately and this analysis can be used in conjunction with the above analysis.

The basic rational for setting priorities of strategic IT investments is nearly as simple as that of Support IT investments. Strategic IT investments with greatest contribution to business objectives that use the least amount of resources should get highest priority. A comparison of benefits to costs can be made for Support IT investments. For Strategic IT investments a comparison of the contribution to business objectives to the costs of the IT can also be made. In some cases the contribution to business objectives may be quantified similar to quantifying benefits of an IT investment. However, it may be more appropriate to judge the strategic contribution as being high, medium or low, just as was done for categorizing IT investments previously discussed. This type of analysis may be used if quantification of the strategic benefit is difficult. Setting priorities for strategic IT investment is not as straight forward as doing so for support priorities; however, a top management team assigned with the evaluation task will need to come to a consensus about which strategic IT investments have the highest strategic contribution compared to costs.

The next category of IT investments is that of factory investment. Setting priorities for factory IT investments with benefits that may be expressed in quantitative terms may be done in the same manner as that for support IT investments. The financial benefits and costs of an IT investment are compared and priorities are set with IT investments possessing the highest ratio of benefits to costs getting the highest priority. However, in some instances, other factors, such as, meeting business objectives, business risk and infrastructure effects, according to Ward (1990), are also important to consider, in addition to the economic factor. In these instances, setting priorities for factory IT investments is much more involved that setting them for strategic and support IT investments. To incorporate factors such as these, a multi-factor scoring method (MFSM), like those presented in Chapter 8 of this textbook may be employed. The priority rule for factory IT investments when considering economic, as well as non-economic factors is as follows: IT investments that score the highest with respect to economic benefit,

CSFs, risk and infrastructure improvement factors and use the least amount of resources should get the highest priority.

Let's illustrate the procedure above with an example. Suppose that an organization must set priorities among three different factory IT investments where both economic and non-economic factors are important. Ten managers have been selected to rate each alternative factory investment on the four factors suggested by Ward (1990) and the average rating of the ten will be used to determine the priority of investment. The three alternatives were rated on a scale of 1 to 9, where 1 represents a "poor rating" in satisfying a criteria and 9 represents a "good rating" in satisfying a criteria. Also suppose that weights have been assigned to each of the factors according to their evaluation importance. The weights for the economic, CSFs, business risk, and infrastructure enhancement factors are 35, 30, 25, and 20 percent, respectively. As shown in Table 2, the average rating is multiplied by the factor weight and then the scores are summed for each alternative investment.

Table 2. Multi-factoring scoring method for factory IT investments.

	Factor Weights	Factory Computer System A	Factory Computer System B	Factory Computer System C
Economic benefit	0.35	0.35 x 9 = 3.15	0.35 x 2 = 0.70	0.35 x 4 = 1.40
Strategic benefit	0.30	0.30 x 2 = 0.60	0.30 x 9 = 2.70	0.30 x 5 = 1.50
Business risk	0.25	0.25 x 5 = 1.25	0.25 x 5 = 1.25	0.25 x 5 = 1.25
Infrastructure enhancement	0.20	0.20 x 8 = 1.60	0.20 x 7 = 1.40	0.20 x 4 = 0.80
Total	1.0	6.60	6.05	4.95

The alternative investment with the highest weighted-factor score is the best alternative and the second best alternative is the one with the second highest score, and so on. Considering the weighted-factor scores only, Factory Computer System A should be given the highest investment priority and, thus, should be undertaken before the other two.

Factory Computer System B should be undertaken second and Factory Computer System C, third, assuming that no other investment opportunities arise. However, no consideration has been given to the cost of the investment.

Let us now suppose that the total cost of Factory Computer System A, B, and C are $2,000,000, $1,200,000 and $1,250,000, respectively. Even though Factory Computer System A has the highest factor-rating score it may not be given the highest investment priority because it costs two times the amount of Factory Computer System B and it may be determined that the difference in their weighted-factor scores is not worth the $1,000,000 cost difference. Cost must also be considered in the analysis and may be done so by calculating a benefit-to-cost ratio. IT investment with the highest ratio may be assigned the highest priority. Table 3 shows such a ratio for each alternative investment and the corresponding priorities.

Table 3. Priorities for a factory IT investment problem.

	Benefit/Cost ratio	Priority
Factory Computer System A	$\dfrac{6.60}{2} = 3.30$	3rd
Factory Computer System B	$\dfrac{6.05}{1.2} = 5.04$	1st
Factory Computer System C	$\dfrac{4.95}{1.25} = 3.96$	2nd

As shown in Table 3, Factory Computer System B is the best alternative and should be given the highest investment priority, while C and A should be given second and third priorities, assuming no other investment opportunities arise. If other opportunities arise, then they should be considered in the analysis.

The last category of investments is that of high potential investments. Setting priorities for high potential investments may be even more difficult than for factory investments with non-economic benefits. High potential investments may be likened to R&D investments that often require quick analysis and implementation of a prototype to determine the viability. Often an analysis of the available resources is conducted and then a decision on how best to employ the resources is undertaken. Resources may be of an economic, technical or organizational nature and include such things as technical skills, employee knowledge, available funds, and organizational culture. Often a project champion emerges with an idea to improve organizational performance. This person evaluates the available IT investments and makes a suggestion to management about the best alternative. It seems that the project champion who makes the best case often receives approval for an investment. A project champion may choose to use any type of evaluation technique to suggest recommendations for investment. CSFs analysis may also be used to support a project champion's case. A factor-rating method could be used to rate factors, like CSFs to evaluate which high potential investments get priority over others. The priority rule for high potential investments may be not be formally stated in most organizations, however, informally, IT investments with the most influential CSFs and champion enthusiasm should get the highest priority. Again, high potential investments need to be analyzed, and a prototype should be implemented quickly to gain a better understanding of their true effects. Whether a champion is attempting to gain approval for a prototype or a full-blown system, some form of evaluation will be needed to determine the priority of investment. Usually, a champion's enthusiasm factor and another factor that determines how well the investment will assist in meeting organizational objectives are used to determine priority.

A summarization of the priority rules for each category of IT investment are shown in Table 4. Again, for each category, when appropriate, benefits, costs, and risk are assessed for each alternative and priorities are set based on analysis of these factors. For some types of investments the economic factors are most important, while other type's non-economic factors are most important and the evaluation techniques selected must account for these differences.

Table 4. Priority rules for each IT investment category.

Investment category	Priority rule
Strategic	IT investments with the greatest contribution to business objectives that use the least amount of resources should get highest priority
High potential	IT investments with the highest value of CSFs and a champion factor should get highest priority
Factory	IT investments which score the highest with respect to economic, CSFs, risk and infrastructure improvement factors and employ the least amount of resources should get highest priority
Support	IT investments with the greatest economic benefit that use the least amount of resources should get highest priority

Investment Mapping Methodology

Investment mapping methodology is a comprehensive evaluation technique that consists of prescriptions for creating an investment portfolio map, evaluating overall IT strategy, identifying and evaluating individual investments, and managing benefits after implementation. This methodology may be used as an overall IT management tool for assessing and monitoring every aspect of IT investment. Investment mapping was developed by Glen Peters (1988) as a method to identify the benefits of an IT investment and to effectively manage those benefits. Traditionally, the benefits of an IT investment have been thought of as tangible and related to some type of cost savings. Tangible benefits are quantifiable and thus tend to be easily measured and monitored. As ITs have advanced and evolved, the benefits of IT investments have also evolved. The objectives of IT investments, and their corresponding benefits, have become less tangible and thus more difficult to measure and manage.

The major benefits of today's IT tend to be intangible benefits such as improved customer satisfaction, higher quality information, and better corporate image. Measuring and managing these types of benefits can be very difficult. Peters (1989) suggests using surrogate measures for intangible benefits because identifying expected benefits, tangible or intangible, and actually measuring their effects, are important aspects of managing IT investments. Investment Mapping may be used to evaluate IT strategy, to align IT strategy with individual IT investments, and to evaluate individual IT investments.

Investment mapping consists of the following three stages as identified by Peters (1989): (1) mapping of the investment portfolio, (2) evaluation of alternative investments and performance measurement, and (3) management of benefits after implementation.

Mapping the investment portfolio

In 1987 Peters (1989) conducted a study to determine how companies identified, evaluated and managed IT investments. He analyzed nearly 30 IT investments at four international organizations in the UK. Through personal interviews he deemed benefits identification and management as important to the success of an IT investment, as well as the relationship between an IT investment and the organization's value chain. In the investment mapping methodology, these two factors are the vertical and horizontal axis of the investment map or portfolio as shown in Figure 1.

Peters (1989) identified three types of benefits that include: (1) enhance productivity, (2) minimize risk, and (3) expand the business. These types of benefits should be viewed as a continuum of possible benefits, with those that enhance productivity at one extreme and those that expand the business at the other. Benefits associated with minimizing risk are in between the two extremes. IT investment benefits that enhance productivity tend to decrease costs in some way, such as to decrease production or warehousing costs. These benefits tend to be easily measured in monetary terms and may be referred to as tangible benefits. IT investment benefits associated with expanding the business are usually associated with creating new ways to do business, to produce products and services, and to solve business problems. A common benefit of expanding the business may be "improved customer satisfaction", which is an intangible benefit that is difficult to measure. IT investment benefits that minimize risk can be realized in numerous areas of the business depending on the actual investment. For example, accounting systems minimize the risk of errors and fraud, while decision support systems decrease the risk of making poor management decisions. Benefits that minimize risk may be either tangible or intangible depending on the actual benefit.

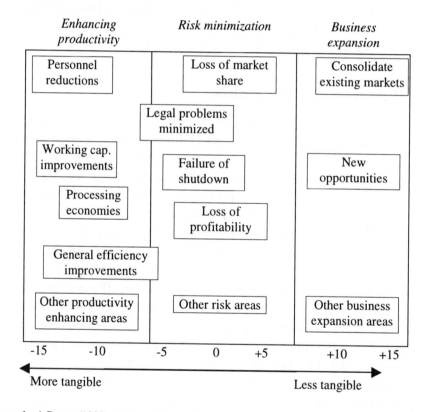

Figure 1. A Peters (1989) type benefits continuum.

The continuum of benefits in Figure 1 assists decision makers in determining the placement of actual IT investments on the investment portfolio map. A scale from -15 to +15 is used for the continuum of benefits, where investments that are more tangible are rated on a scale of 0 to –15 and those that are less tangible are rated on a scale of 0 to +15. It should be noted that the negative points do not represent negative benefits but are only used for plotting investments on a portfolio map. Notice in Figure 1, that each main category of benefits has a scale of 10 points and that a total of 30 points covers the entire scale. Also notice that more specific benefits, subcategories, are identified for each main category (e.g., personnel reductions and processing economies are subcategories of benefits that enhance productivity). In the evaluation process, decision makers would identify the specific, major benefits of an IT investment and then determine the tangibility score. In Figure 1, an investment with "personnel reductions" as the major benefit would be plotted between –15 and –11 on the vertical axis of the investment portfolio map. Decision makers must decide on the size of the box depicting the benefits on the benefits continuum, as the size represents the degree of tangibility of an investment benefit. In the event that major benefits are in more than one main category of investment benefits, decision makers may select the most important one and use it to plot the investment. Alternatively, decision makers may decide that all benefits are important and plot the IT investment across all main categories.

Once benefits have been identified and scored, the position of the IT investment with respect to its orientation with the value chain of an organization must be determined. Peters (1989) identified three categories of investments that represent the orientation of investments. The three categories of investments are: (1) support the technical infrastructure, (2) perform routine business operations, and (3) influence the market. IT investments that support the technical infrastructure are those investments in telecommunications, processors, software environments, and shared applications. Technical infrastructure investments provide the base for which all other IT investments are to be operationalized. Peters (1989) suggests that infrastructure investments should be considered individually as opposed to being included in an actual information system. Infrastructure investments need to be considered separately to take into account their individual importance and their importance to future IT investments. IT investments that perform routine business operations support the regular business processes of an organization. Examples of these types of systems include order-processing systems, scheduling systems, office systems, finance and accounting systems and logistics systems. IT investments that influence the market are those that assist in changing the buying patterns of customers. These types of IT create new distribution channels and may bring products and services closer to the customer in some way. The Internet and IT investments necessary for e-commerce are examples of market influencing investments.

Just as the benefits of an IT are viewed as a continuum, the types of IT investments may be viewed as a continuum of investments. The investment orientation continuum is presented in Figure 2.

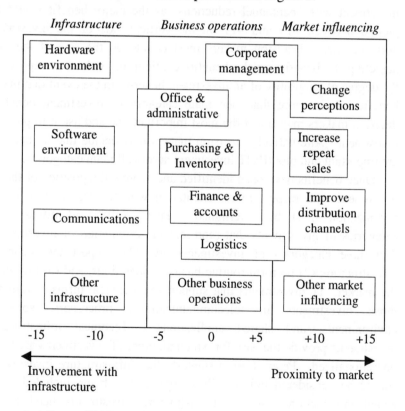

Figure 2. A Peters (1989) type investment orientation continuum.

The investment orientation continuum in Figure 2 is used to determine the horizontal position of individual IT investments on the investment portfolio map. The same scale of -15 to +15 is used to score the investment orientation of an IT investment. Each category of investment orientation has 10 points for a total of 30 points for investment orientation. IT investments that are highly related to the technical infrastructure will be scored –15 to -5, while those close in proximity to the customer will be scored somewhere between +5 and +15. IT investments that support the business operations of an organization will be scored between –5 and +5. Notice that in Figure 2, some of the IT investments, like those of "logistics" and "office & administration" cross two types of IT investments. It is reasonable to accept that a "logistics" system may both influence customers and support business processes, while an "office & administration" system both supports business processes and is part of the technical infrastructure. IT investments that cross into different categories of investments, as well as benefits are quite common in Investment Mapping.

Evaluating IT investments

As mentioned before, Investment Mapping may be used to evaluate IT strategy, to align IT strategy with individual IT investments, and to evaluate individual IT investments. In each case an investment portfolio map must be constructed. The map is constructed by plotting individual IT investments using the coordinates for the vertical axis as the type of benefits, and those for the horizontal axis as the type of investment. Figure 3 presents the investment portfolio map.

Decision makers evaluate an IT investment and determine the "tangibility" score for benefits and the score that reflects the type IT investment or its orientation. These scores are then used to plot the individual IT investment. Suppose an organization is trying to determine whether or not to invest in *electronic data interchange* (EDI) system to better connect to suppliers. The group has evaluated the benefits of such a system and determined them to be mostly tangible and scored them as

–10 to 3.   The benefits in this case both enhance productivity and minimize risk for the organization.   Decision makers have also determined the type of IT investment to be mostly a technical infrastructure investment but also one that to some extent, supports business processes. As a result, the investment is scored as –10 to –4 for investment orientation.

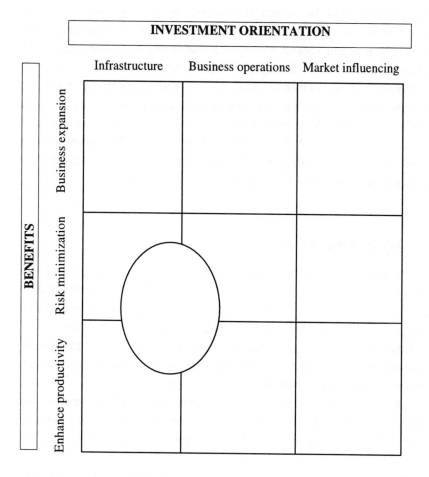

Figure 3. Investment portfolio map.

Using these coordinates, the IT investment could be plotted as an oval on the investment portfolio grid, like the one presented in Figure 3. A complete investment portfolio map must be constructed that contains each individual IT investment necessary for the type of evaluation that it is intended to support.

Once a complete investment portfolio map has been constructed it may be used to assist decision-making for its intended purpose. Decision makers wanting to evaluate IT strategy will plot a map of IT investments and evaluate whether or not the IT investments support overall corporate strategy and then will compare such an investment portfolio map to that of competitors. The investment portfolio map may also be used to determine if an individual IT investment "fits" with the overall business strategy of the organization. For a set of alternative investments, the investment portfolio allows decision makers to determine how individual investments fit into the overall portfolio and subjectively, determine which fits best. Further, decision makers may choose some other evaluation technique like cost/benefit analysis, net present value analysis or payback period to complement investment portfolio methodologies. The investment portfolio may be utilized to ensure a set of alternative investments fit equally well into the portfolio of investments and then a supplemental technique, like cost/benefit analysis, may be used to determine which alternative is best.

Because investment mapping methodology is focused on benefits identification and management, Peters (1989) developed a framework called the *cost-benefit hierarchy* that assists decision-makers in selecting measurable metrics to evaluate IT investment benefits. The cost-benefit hierarchy is a framework that may be used to value the benefits of an IT investment. The value of benefits is based on their expected impact on profitability. In this framework, it is assumed that there are two types of benefits that impact profitability, those that provide a cost savings and those that generate revenue. The cost-benefit hierarchy depicts the costs savings variables and their impact on profitability on one side of the hierarchy and revenue generating variables and their corresponding effects on the other side.

The first step in developing a cost-benefit hierarchy is for users, or some other group of stakeholders, to identify key measurable variables that reflect the impact of expected benefits on organizational performance (examples provided in Chapter 1). At least one key variable should be selected to measure each major benefit. Measurable variables are ones that can be measured in quantitative terms and possibly, but not essential, in monetary terms. Some examples of measurable variables are inventory turnover, customer waiting time, time to make repairs, and number of sales calls per day. It is assumed that all benefits, whether tangible or intangible can be measured in quantitative terms. Intangibles benefits must be measured with surrogate variables that decision makers deem as the most appropriate surrogate measures. The key measurable variables are depicted as the base in the cost-benefit hierarchy.

The next step in developing a cost-benefit hierarchy is to identify the expected changes in key variables and their impact on profitability. The expected changes and effects are depicted on the second level of the hierarchy, while impacts on profitability are depicted at the top level as shown in Figure 4.

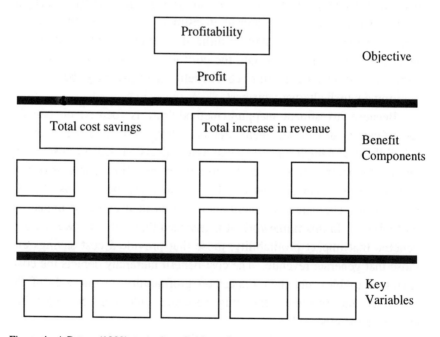

Figure 4. A Peters (1989) cost—benefit hierarchy.

Decision makers discuss the possible changes of each key variables individually and determine their individual impact on firm profitability. Suppose they must assess the impact of "increased customer loyalty" on firm profitability. They may decide that the variable "sales" is the most appropriate surrogate measure for "increased customer loyalty" and the IT investment will result in a five percent increase in sales. The five percent increase in sales may be translated into a dollar figure representing the monetary impact on profitability. Once each key variable's impact has been determined, an overall amount may be assessed for the impact on profitability (i.e., the individual dollar amounts are added together to attain the overall expected increase in profitability). The expected increase in profitability may be used as the overall value of benefits attributable to the IT investments. This value may be used to evaluate alternative IT investments. The value may also be used as input into other types of analysis, such as in net present value or payback period analysis.

Managing benefits

The third and last stage of the investment mapping methodology is to manage the benefits of an IT so that they actually materialize favorably for an organization. Peters (1989) suggests assigning responsibility to managers for reaching the key variable targets defined when creating the cost-benefit hierarchy. In the example above, it was assumed that an IT investment would increase "customer loyalty" and in turn increase sales by five percent. A manager would be assigned responsibility to use the IT to improve customer loyalty so that sales would increase by five percent. Managers then assign support responsibilities to subordinates to actually bring about the five percent increase in sales. A manager or a group of managers are assigned responsibility to manage each of the key variables and achieve their target or expected effects. Assigning responsibility may be an appropriate method of managing benefits; however, its use is cautioned. In many cases the key variables are surrogates for major benefits of an IT and these key variables may or may not be good measures. Great care must be taken in selecting surrogate measures so that they truly measure what they are intended to measure and managers are focusing on factors that make a difference. Further, the subjectivity that decision makers are required to use in many stages of the investment mapping methodology may cause concern. Consequently, organizations that use this methodology should be aware of its disadvantages and be prepared to compensate for them.

Other Investment Portfolio Methodologies

Other investment portfolio methodologies exist in the literature. Two specific examples of investment portfolio methodologies are the Berghout and Meertens (1992) investment portfolio methodology and Bedell's (1985) methodology (Van Reeken, 1992). The Berghout and Meertens (1992) investment portfolio methodology evaluates an IT investment's contribution to the business domain, contribution to the technology domain and the financial effects with net present value analysis. To assess the business and technology contribution, the evaluation criteria of the information economics methodology are used in a "weighted-scoring method" similar to those presented in Chapter 8 of this textbook. IT investments are plotted on a map with contribution to business domain on the vertical axis and contribution to technological domain on horizontal axis. The financial analysis of net present value determines the size of the circle on the map that represents individual IT investments.

Bedell's methodology assumes that IT supports business activities and some business activities are more important than others. It also assumes that the effectiveness and efficiency of an IT define its quality. IT investments are prioritized with a total contribution score calculated as the score for importance of an IT multiplied by the score for quality of an IT. Three portfolios are established to show the (1) overall investment strategy, (2) the investment areas of all business activities, and (3) investment areas of a particular activity. Importance of an IT investment represents the vertical axis, while quality of an IT investment represents the horizontal axis.

It should be noted that variations of the methodologies presented in this chapter might be employed by organizations so to as to customize the techniques to support their specific organizational situations and use. Organizations may also develop their own portfolio methodologies. These applications of portfolio management may be as simple as databases containing cost, benefit and risk information for each IT investment or may be very complicated mathematical models, based on Markowitz's modern portfolio theory, used to evaluate IT investments. See Markowitz (1970) for a detailed discussion of modern portfolio theory.

Summary

We introduced investment portfolio methods for IT planning. Two differing types of portfolio methods were presented. Ward's (1990) method categorizes IT investments by the role they play in the organization and the expected contribution they will make to business performance. Peters (1989) investment mapping method was developed to identify the benefits of an IT investment and to effectively manage those benefits. Both methods are flexible to allow customization in their application in unique circumstances found in differing businesses. The preponderance of references in this chapter to methodologies in other chapters attests to the ability permitted decision makers to incorporate their preferred methodologies within the context of portfolio methods.

In the next chapter we examine the use of survey methods to achieve IT investment planning information. Like this chapter, the next chapter's methodologies can utilize many of the previously mentioned methodologies as a collection or portfolio of methodologies useful as an aid in IT decision-making.

Review Terms

Champion
Cost-benefit hierarchy
Critical success factors (CSFs)
Electronic data interchange (EDI)

Investment mapping methodology
Investment portfolio methodologies
Portfolio management
Setting priorities

Factory investments	Strategic investments
Feasibility study	Support investments
High potential investments	Ward's portfolio approach

Discussion Questions

1. How do portfolio methods differ from basic financial methods, like return on investment? Do they really differ?
2. Where do we use a "champion" in the Ward's methodology? Why do we need the "champion"?
3. A "feasibility study" is needed in the Ward methodology. Why?
4. Why is setting priorities useful in IT investment decision-making when using the Ward's methodology?
5. Is investment mapping just a graphic exercise or does it help in planning IT decisions? How?
6. Why do we have to have both a "benefits" and an "orientation" continuum in the Peter's methodology?
7. What does the scoring from -15 to +15 try to accomplish in the Peter's methodology?
8. Explain the use of the "cost-benefit hierarchy" in the Peter's methodology?

Concept Questions

1. If you had to define portfolio methods in a single sentence, what would it be?
2. Why in the Ward (1990) methodology do we have to categorize IT investments by their role in the organization and their expected contribution to business performance?
3. How do "strategic investments" differ from "high potential investments" in the Ward methodology?
4. How do "factory investments" differ from "support investments" in the Ward methodology?
5. How do we use "critical success factors" in the Ward's methodology?
6. How are we able to use multi-factoring scoring method for IT investments?

7. How does investment mapping methodology help identify the benefits of an IT investment and to effectively manage those benefits?
8. What are we trying to identify with the three categories of "enhancing productivity", "risk minimization" and "business expansion" in the Peter's methodology?

References

Bedell, E.F., *The Computer Solution: Strategies for Success in the Information Age*, Homewood, Dow-Jones Irwin, 1985.

Berghout, E.W. and Meertens, F.J.J, "Investment Portfolio for the Evaluation of IT Investment Proposals," (in Dutch), *Informatie*, 1992, pp. 677-691.

McNurlin, B.C. and Sprague, R.H., *Information Systems Management In Practice*, 6th ed., Upper Saddle River, NJ: Pearson/Prentice Hall, 2004.

Markowitz, Harry, *Portfolio Selection: Efficient Diversification of Investments*, Cowles Foundation Monograph 16, Cowles Foundation for Research in Economics at Yale University, 1970.

Peters, Glen, "Evaluating Your Computer Investment Strategy," *Journal of Information Technology*, September 1988, pp. 123-134.

Peters, Glen, The Evaluation of Information Technology Projects, PhD Thesis, Brunel University, 1989.

Peters, Glen, "Beyond Strategy: Benefits Identification and Management of Specific IT Investments," *Journal of Information Technology*, 1990, Vol. 5, pp. 205-214.

Van Reeken, A.J., *Selection of Investments in Information Systems*, Eugene Bedell's Method, (In Dutch), Handbook BIK, 1992, pp. 1030-1032.

Ward, John M., "A Portfolio Approach to Evaluating Information Systems Investments and Setting Priorities," *Journal of Information Technology*, 1990, Vol. 5, pp. 222-231.

Chapter 12

Value Analysis and Benefit/Risk
Methodologies

Learning Objectives

After completing this chapter, you should be able to:

- Explain what "value analysis methodology" is and how it can be used in IT investment decision-making.
- Describe the steps in a "value analysis".
- Explain what "Benefit/risk analysis methodology" is and how it can be used for in IT investment decision-making.
- Explain how "risk assessment" questions can be used in IT investment decision-making.

Introduction

In this chapter we examine two methodologies that can combine qualitative and quantitative measures of IT investment performance by utilizing survey methods. These methodologies include "value analysis" and "benefit/risk analysis". Both can utilize, and build on many of the previous methodologies presented in this textbook.

What is Value Analysis Methodology?

Value analysis methodology is an IT investment evaluation technique that involves prototyping and surveying to determine the value of benefits of an IT investment (Chase et al, 2004, pp. 161-162). Many information systems evolve and change throughout their lives according to user need. Value analysis may be utilized initially at the evaluation and selection stage. It can also be used each time the system goes through one of these changes or evolutions. Value analysis begins with proposing an IT prototype and determining its costs. An *IT prototype* can be viewed as a preliminary version of the IT or information system that is being planned. It is not a full-working version of the desired system, just a "make-do" system to allow for comparison. Then potential benefits of an IT prototype are identified and valued. Survey techniques may be used to determine the value of the identified benefits. Once a value has been assigned to the benefits, this value is compared to the cost of the IT prototype. If the value of benefits exceeds the costs of the IT prototype, then the prototype should be built. If not, then the project may be scaled-down to reduce costs and then re-examined. Several iterations or evolutions of value analysis may take place either initially or throughout the life of the information system and its use of the IT.

Value analysis methodology may be compared to a feasibility study in the systems development lifecycle. The *systems development lifecycle* is the traditional systems development method utilized at the present time to build an information system. It consists of a set of processes that includes investigation, systems analysis, systems design, programming, testing, implementation, operation and maintenance. During the first process of investigation, a feasibility study is conducted to determine technical, operational and behavioral feasibility. The use of prototyping in the value analysis methodology accomplishes the same things as a feasibility study (Kangas, 2003; Keen, 1981; Laudon and Laudon, 2004, pp. 395-397). Prototyping allows for the determination of whether or not the requirements of the system may be accomplished with existing and available technologies, hence assessing technical feasibility. Operational feasibility may also be assessed through creating and testing a prototype.

Prototyping allows the analyst to determine whether or not the proposed IT prototype will actually provide benefits and improve productivity in such a way as to improve overall organization performance. Behavioral feasibility is most easily demonstrated by the use of prototyping. Actually testing the prototype on users and soliciting their opinions allows for the determination of this type of feasibility.

Value analysis methodology is based on making value-costs assessments of proposed IT investments. Value analysis may be appropriate for IT systems that are associated with major benefits that are qualitative and not easily or appropriately measured by surrogates. Value analysis is also appropriately used in situations where the IT evolves because the user may not know what features the IT should have and only after experimentation with a prototype are users able to identify wanted features and capabilities. Value analysis is an appropriate evaluation technique for innovative IT investments. Innovative IT investments tend to have many intangible benefits that are considered to be the major benefits of the IT. Value analysis methodology provides a framework to incorporate these important intangibles into the analysis. Value analysis methodology may be used to evaluate independent IT investments and to select one or several from a set of alternative IT investments. Value analysis was originally proposed by Keen (1981) as a technique to evaluate *decision support systems* (DSS); however, the technique may be applied to any situation where major benefits are qualitative, the systems evolve as user needs are identified and evaluated, and where traditional cost/benefit analysis seems inappropriate.

Value analysis methodology may be broken down into twelve general steps:

1. Propose IT prototype and determine its cost;
2. Identify and value benefits of prototype;
3. Compare value of benefits with cost of prototype;
4. Design and build prototype;
5. Test prototype and measure its use;
6. Propose new prototype and determine its cost;
7. Identify benefits of new prototype;
8. Value benefits of new prototype;

9. Compare value of benefits with cost of new prototype;
10. Redesign and build new prototype;
11. Test new prototype and measure its use and costs; and
12. Repeat steps 6 thru 11 as necessary to satisfy stakeholder requirements.

The first step in value analysis is to propose an IT prototype and to determine the cost of designing, building and testing it. Once the IT prototype has been proposed, its benefits must be identified and valued. Identifying and valuing benefits of an IT is one of the most important and difficult steps in IT investment decision-making. Typically, a short list of the main benefits of an IT is compiled to be used during the valuation process. To value these benefits stakeholders, most often users, are asked if the benefits of the proposed IT are worth the cost of the prototype or how much they would be willing to pay for the benefits. More specifically, users are asked "Would you be willing to pay $X to attain the benefits of the proposed IT? or "How much would you be willing to pay for the benefits of the proposed IT?" If the value of the IT prototype is greater than its cost, then the IT prototype should be designed and built. When the prototype is complete, testing or experimenting occurs where users actually operate the IT. The prototype needs to be built such that it allows users to experience the actual performance of the IT. As users experiment with the IT, analysts measure the use and take note of the user suggestions and requirements. Based on the use and these suggestions and requirements, analysts propose a new prototype and the steps in the value analysis methodology begin again. The new prototype maybe a completely new system, or more likely a system modified from the original one.

After proposing a new IT prototype, benefits are identified and valued. Often, the list of benefits for the original IT prototype is expanded or modified to reflect the changes in the IT to create a new list of benefits. In other cases the benefits list remains unchanged and may be used for evaluating the new IT prototype. The new benefits are valued using the same survey technique as before. If the value of benefits exceeds the costs of the new IT prototype, then it will be built and tested. The value analysis methodology may repeat as many times as necessary and whenever stakeholders request it.

Let's conceptually illustrate the use of the value analysis procedure. Suppose an organization is deciding whether or not to invest in a DSS. It has been determined that a prototype of the proposed DSS will cost $10,000 to design, build and test. A list of the expected benefits of the proposed DSS includes the ability to examine more alternatives, improve communication of analysis, make better decisions, provide flexible report generation, and stimulate new ideas. Note the benefits associated with a DSS tend to be intangible, meaning they cannot readily be assigned a value measure, like dollars. Indeed, assigning an accurate value to an intangible, such as "improved communication of analysis", would be very difficult, if not impossible. Value analysis methodology allows decision makers to determine the value of intangible benefits in such a way as to reduce error in estimation. Survey techniques such as the Delphi method (from Chapter 7) may be employed to determine the value of benefits.

The Delphi method has been proposed as an evaluation technique to assess the benefits of a system by Powell (1992), and Parker, Benson and Trainor (1988). Farbey, *et al.* (1994) suggests experiment/role playing, management game or simulation in addition to the Delphi technique to assess benefits of an IT investment. Let's say here that in the first round of the Delphi method, forty users were given the list benefits and then asked how much they would be willing to pay for them. The results were complied and in a second round, first round feedback was given to users. The users were asked again to estimate the value of the proposed benefits given the first round results. A third round is then conducted and, for purposes of this example, it was agreed by all evaluators that the value of benefits of the DSS is $15,550. In some cases the number of benefits may be rather large compared to the number in this example. In these cases, statistical analyses may be conducted (e.g., cluster analysis, to group the benefits assigned to the groups rather than to individual benefits).

Continuing with the case situation, the organization should build and test the prototype, as the cost of the prototype, $10,000, is less than the value of the benefits, $15,550. After implementing and testing the prototype, analysts took the actual usage, suggestions and requirements of users to propose a newly modified DSS. This DSS was found to be cost justified. After several more iterations of the value analysis methodology, it was determined that that no further improvements to the DSS should be made and the process ended. Sometime in the future, as stakeholders deem it necessary, the value analysis process may be used again to improve and enhance the DSS.

Advantages of the value analysis methodology are that the technique emphasizes the value of benefits over costs and the evaluation of costs and benefits are kept separate. Another advantage is that value analysis considers intangible benefits of IT investments in such a way that surrogate measures do not have to be defined and measured. The intangible benefits are incorporated into the analysis by asking users the amount they would be willing to pay for the benefit. Another advantage is that prototyping accompanying the value analysis methodology decreases risk associated with investing in IT. *Risk* can be defined here as a possibility that an event will cause an organization to fail to meet its goals (Gelinas *et al.*, 2004, p 243). A prototype costs less than a full system and thus, the size of the project is smaller, resulting in less risk for the organization. With the use of prototyping, an IT system may be seen as a *research and development* (R&D) investment rather than a capital investment. As such, investments in R&D can be written off if the system fails, and risk is minimized further. The main disadvantage is that value analysis should not be used for investments that are large and considered to be capital investments. Keen (1981) suggests that the cost of the original prototype should not exceed $20,000.   Capital investments, those greater than $20,000 (in 1981 dollars) should be assessed with more traditional capital budgeting techniques like net present value analysis, return on investment and internal rate of return. Value analysis is an appropriate evaluation technique when the main objective of the IT investment is to improve future effectiveness of an organization.

What is Benefit/Risk Analysis Methodology?

Benefit/risk analysis methodology is an IT investment evaluation technique in which the benefits and risks of an IT investment are assessed and then compared to determine if the benefits outweigh the risks. Many IT investment projects are late, over budget or cancelled. McFarlan and McKenney (1983) contend that one of the major reasons for this phenomenon is that organizations do not assess the risks of their IT investments. Not assessing risk may cause an IT investment to fail to obtain some or all of the anticipated benefits, incur more costs than originally expected, and take longer for implementation. In addition, the IT may not perform as expected and may be incompatible with other IT. In many situations, assessing the risk of an IT investment may lead to higher-quality IT investment decisions than without the assessment of risk. Other mainline IT authors claim risk assessment is a fundamental component of any organization information systems control structure (Kangas, 2003; Laudon and Laudon, 2004, p. 464).

Most of the methodologies presented in this textbook have the capability to either explicitly or implicitly considere risk. However, the benefit/risk methodology provides a framework that may be used to explicitly examine risk and assess its impact on IT investment decision-making. By assessing risks, as well as benefits and costs of an IT investment, the organization may realize benefits to the full extent and, thus, experience improvements in overall organizational performance.

Benefit/risk analysis involves first identifying possible benefits and risks of an IT investment and then assessing their effects (McNurlin and Sprague, 2004, pp. 396-402). Benefits identification and quantification are usually conducted in a feasibility study performed by the organization before implementation of a new IT. A feasibility study usually results in information such as qualitative and quantitative benefits, costs, completion target dates and staffing needs. Consequently, extensive effort is usually made to assess benefits and costs of an IT investment. Benefit assessment may be conducted using any of the techniques that provide a "benefits-assessment", such as those in cost/benefit analysis (Chapter 6) and the investment mapping methodology (Chapter 11). In both, cost/benefit analysis and the investment mapping methodology, benefits are identified and their overall impact is quantified, and presented in monetary terms, when possible. Benefit/risk analysis focuses on the benefits and risks of an IT investment; however, costs should also be considered. Cost analysis may be performed during a feasibility study with a technique like cost/benefit analysis. An *IT team* or project champions are usually charged with the task of identifying and assessing benefits, costs and risks and presenting their recommendations to final decision makers for IT investment approval.

Once benefits have been identified and assessed, then risk must be taken into consideration. We assume that management has used the appropriate tools and methods to evaluate and manage IT investments and that risk is what remains after use of these proper tools. Many firms undertake a formal *risk assessment*, which determine points of vulnerability in information systems (Laudon and Laudon, 2004, p. 464). Several dimensions of risk influence the inherent risk of an IT investment. McFarlan and McKenney (1983) identify the following three dimensions of risk: (1) project size; (2) experience with technology; and (3) project structure. Project size may be defined in terms of the total dollar cost of the IT project, time to fully implement an IT investment, number of staff needed for the project, or number of departments affected by the investment. The larger these factors are, the higher the risk of an IT investment. The risk associated with project size is relative to the experience an organization has in developing and implementing a particular sized project. An organization that is more experienced at developing and implementing larger sized projects has less risk with larger IT investments than an organization with less experience. The second dimension of risk is related to the organization's experience with the technology. An organization with extensive experience with technology has less IT investment risk than another with little experience. Experience with technology involves the degree of familiarity people in the organization have with the technology. The third dimension of risk is the structure of the IT investment. Highly structured investments have easily defined outputs that do not change during the life of the investment, and, thus, have less risk. Less structured investments are ones that decision makers have difficulty defining the outputs of the investment and the definition may change during the life of the investment. These dimensions of risk, project size, experience with technology and project structure, were combined and presented in a grid like the one in Figure 1.

	High structure	Low structure
High experience with technology	Large size—low risk	Large size—low risk
	Small size—very low risk	Small size—very low
High experience with technology	Large size—medium risk	Large size—very high risk
	Small size—medium-low risk	Small size—high risk

Figure 1. A McFarlan and McKenney (1983) type project risk grid.

The horizontal axis of the grid shows project structure and the vertical axis represents the organization's experience with the technology. Size of project is presented within each cell, as well as the overall project risk. Notice that when the organization is experienced with the technology, "High experience technology", the risk of the project is considered to be low regardless of project size and structure. Also notice that overall project risk increases as the risk of the three individual dimensions increases. Analyzing individual IT investment risks with this grid can improve decision-makers' understanding of project risk and the grid can be used as a management communication tool.

To further assess the risk of IT investments, McFarlan and McKenney (1983) and Rainer *et al.* (1991) suggest using a *risk assessment questionnaire.* A scoring and weighting method can be incorporated into the risk assessment questionnaire to aid in comparisons on differing projects. Tables 1, 2 and 3 contains sample questions that illustrate a risk questionnaire based on McFarlan and McKenney (1983). This particular questionnaire was designed it so that an overall risk score could be calculated for individual IT investments. The higher the score the more risk an IT investment has. Notice that each question is assigned a weight according to its importance in assessing risk. Also notice that questions were developed to measure risk in each of the three dimensions of risk, which include project size, experience with the technology, and project structure.

Table 1. Sample of "size" risk assessment questions.

Questions	Score	Weight
1. Total development IT staff hours for system		10
100-3,000	Low-1	
3,000-15,000	Medium-2	
More than 15,000	High-3	
2. What is the IT project implementation time?		20
12 months or less	Low-1	
13 months or more	High-3	

Organizations should develop risk questionnaires that are tailored to their needs and the type of technology under consideration. The risk assessment questionnaire may be used at various times during evaluation and implementation, as considered necessary. Under normal circumstances, the risk of an IT investment should decrease as implementation progresses favorably, and thus, the risk scores of subsequent surveys should also decrease. As implementation progresses, users and IT professionals become familiar with the technology and more experienced, so that IT investment risk decreases.

Table 2. Sample of "structure" risk assessment questions.

Questions	Score	Weight
1. What is the severity of process changes in the user department that will be caused by the proposed system?		8
Few changes	Low-1	
Moderate changes	Medium-2	
Extensive changes	High-3	
2. Does user organization have to change structurally to meet requirements of new system when it is in operation?		12
No	-0	
Little	Low-1	
Somewhat	Medium-2	
Much	High-3	

Once the risk assessment questionnaire has been distributed and results analyzed, the benefits of an IT investment should be compared to the risks of such an investment. Decision makers subjectively analyze whether or not the benefits are worth the risks of an IT investment. It is assumed that higher-risk projects must yield higher benefits to compensate for more risk. Decision makers may ask questions like the following to make this assessment (McFarlan and McKenney, 1983):

1. Are the benefits large enough in comparison to the risks?
2. Can affected parts of the organization survive if the project fails?
3. Have appropriate alternative IT investments been considered?

Although benefit/risk analysis is a subjective methodology it explicitly considers the risk of an IT investment. This main advantage makes benefit/risk analysis an attractive technique that may be used in conjunction with other techniques so that benefits, risks and costs are each considered in evaluation. McFarlan and McKenney (1983) suggest using the same risk assessment technique to evaluate the overall risk of all IT investments. A new risk questionnaire may be developed to assess the overall risk and determine whether or not the organization has the "right" amount of risk (Curley and Henderson, 1992). In some situations organizations must invest in risky projects to remain competitive. In these situations the overall risk should be higher than that of an organization that does not rely as much on technology to sustain competitive advantage. In any case, organizations must decide on the appropriate level of overall risk and make adjustments to their IT investment strategy to meet that level.

Table 3. Sample of "technology" risk assessment questions.

Questions	Score	Weight
1. Which of the IT hardware is new?		10
None	-0	
CPU	High-3	
Peripheral and/or additional storage	High-3	
Terminals	High-3	
Other-Specify	High-3	
2. Is the IT software new?		10
No	-0	
Programming language	High-3	
Database	High-3	
Data communications	High-3	
Other-Specify	High-3	
3. How knowledgeable is user in the area of IT?		10
First exposure	High-3	
Pervious exposure, limited knowledge	Medium-2	
High degree of exposure	Low-1	
4. How knowledgeable is IT staff in proposed application area?		10
Limited	High-3	
Understands concept but no experience	Medium-2	
Has prior experience	Low-1	

Table 4. Delivery and enablement risk questions based on Currie (2003) performance indictors.

Key performance indicator	Example of risk question
Service availability	What time frame of delivery can the vendor guarantee?
Delivery of end-to-end solution	How will software changeover downtime affect the business? Will business processes need to be changed?
Ability to accommodate changes	How quickly can vendor provide more "seats" if business grows?
Ability to transfer existing data	Can vendor incorporate existing databases into the software application?
Data security and integrity	Does vendor have good data security and integrity system? Does vendor own its own data center?
Disaster recovery, back-up and restore procedures	How quickly will data be recovered if a disaster happens?

As a supplement to the risk assessment questionnaire presented above, another framework has been developed by Currie (2003), to assess the risk associated with outsourcing web-enabled software applications. Outsourcing has become a normal activity for many organizations, especially the outsourcing of IT tasks (note discussion in Chapter 2). Currie's (2003) framework contains a set of questions that may be used to measure the risk of outsourcing the specific task of web-enabled applications. The framework may be used as an example so that an organization can create their own questions to evaluate an *application service provider* (ASP) or any other IT vendor. Currie (2003) identified key performance indicators and corresponding risk questions for the following five categories of vendor performance: (1) delivery and enablement, (2) integration, (3) management and operations, (4) business transformation and (5) client/vendor relationship. Tables 4, 5 and 6 present some these key performance indicators and selected questions to assess the risk of outsourcing based on Currie (2003), Chiesa *et al.* (2000), Earl (1999), and Greaver (1999).

Table 5. Management and operations risk questions based on Currie (2003) performance indictors.

Key performance indicator	Example of risk question
Reduce cost of ownership	How does the cost of outsourced delivery compare with in-house delivery?
Eliminate the problem of managing IT	How can the vendor save effort in managing IT?
Gain access to IT skills	Who will manage the software application contract at the vendor site?
Achieve greater "visibility" of IT costs	How does the cost of providing software applications in-house compare with an outsourced resource?
Improve customer service	What will the customer gain from outsourced software application?
External more cost-effective than traditional outsourcing	How much does in-house software application cost to run?
	How much does outsourced software application cost to run?
	How can we leverage advantages in licensing and maintenance costs with the vendor?
	How can business avoid costs creeping upwards after the contract is signed?
Greater flexibility of outsourcing software applications	What are the intangible benefits of software application when outsourcing?
	What are the tangible benefits of software application when outsourcing?

Table 6. Client/vendor relationship risk questions based on Currie (2003) performance indictors.

Key performance indicator	Example of risk question
Desire to develop partnerships	How can we leverage our relationship with the vendor?
Outsourcing success depends on good service level agreements (SLA)	What is the vendor's SLA offer? What are the industry standards for SLAs? Can the vendor offer customer references?
Financial stability of vendor	How financially survivable is the vendor? How do discontinue our relationship if the vendor goes out of business? What will happen if strategic alliances/partnerships break down between vendors?
Vendor relationship	How can trust be developed with vendor? How can software applications be managed in a climate of economic uncertainty?

Summary

In this chapter we have examined the use of two survey methods: value analysis and benefit/risk analysis. Value analysis is a methodology based on prototyping. Benefit/risk analysis on the other hand directly seeks to use the opinions of users to determine and compare benefits and risks in using IT. Both methods can make use of other financial and non-financial decision-making methods to aid IT investments planning.

Throughout this textbook we have sought to explain, illustrate, and demonstrate a variety of differing and highly practical methods for IT investment decision-making. We have consistently recommended that it takes more than one methodology to render a "good" decision. Indeed, in Chapter 11 we introduced "portfolio investment methodology" which is itself, an advocacy of using multiple IT investment methodologies in concert to attempt to arrive at a "good" IT investment decision. But what if you don't make the "right" decision? To attempt to answer that question and make some final suggestions on strategies for avoiding this situation we end this textbook with an epilogue chapter entitled, "The Costs of Not Making the Right Information Technology Investment Decision (and Strategies on How to Avoid Them)".

Review Terms

Application service provider (ASP)	Research and development (R&D)
Benefit/risk analysis methodology	Risk assessment
Decision support systems (DSS)	Risk assessment questionnaire
IT prototype	Systems development lifecycle
IT team	Value analysis methodology

Discussion Questions

1. Why is prototyping so important in value analysis methodology?
2. Where might value analysis methodology not be useful in planning?
3. Since both value analysis and benefit/risk analysis use survey methods, how are they different?
4. Do all risk assessment questionnaires look alike or are they different?
5. Why might it be necessary to use weightings and scores in risk assessment questionnaires?

Concept Questions

1. What are the twelve steps in value analysis methodology?
2. Not assessing risk may cause an IT investment to what? Explain.
3. If the benefits and risks of an IT investment are assessed and then compared to determine that the benefits don't outweigh the risks, what are we suppose to do under benefit/risk analysis?
4. According to the McFarlan and McKenney (1983) type project risk grid, if a have a high experience with the technology that we want to invest in, there is high structure to the project we are

planning, and it is large in size and scope, what kind of risk assessment can be expect?

5. What kind of questions would you expect to find on a risk assessment questionnaire? Give examples.

References

Chase, R.B., Jacobs, F.R. and Aquilano, N.J., *Operations Management for Competitive Advantage*, 10th ed., Boston, MA: McGraw-Hill, 2004.

Chiesa, V., Manzini, R. and Tecilla, F., "Selecting Sourcing Strategies for Technological Innovation: An Empirical Case Study," *International Journal of Operations and Production Management*, Vol. 20, No. 9, 2000, pp. 1017-1037.

Curley, K. and Henderson, J., "Assessing the Value of a Corporatewide Human Resource Information System: A Case Study," *Journal of Management Information Systems*, Vol. 9, 1992, pp. 45-60.

Currie, W.L., "A Knowledge-based Risk Assessment Framework for Evaluating Web-enabled Application Outsourcing Projects", *International Journal of Project Management*, 21, 2003, pp. 207-217.

Earl, M.J., "The Risks of Outsourcing IT," *Sloan Management Review*, Vol. 37, No. 3, 1999, pp. 26-32.

Farbey, B., Land, F., and Targett, D., *How To Assess Your IT Investment. A Study of Methods and Practice*, Oxford, Butterworth-Heinemann, 1994.

Gelinas, U.J., Sutton, S.G. and Fedorowicz, J., *Business Processes & Information Technology*, Mason, OH: Thomson/South-Western, 2004.

Greaver, M.F., *Strategic Outsourcing*, New York: NY, AMA Publication, 1999.

Kangas, K., *Business Strategies for Information Technology Management*, Hershey, PA: Idea Group Publishing, 2003.

Keen, P.G.W., "Value Analysis: Justifying Decision Support Systems", *MIS Quarterly*, March 1981, pp. 1-15.

Laudon, K.C. and Laudon, J.P., *Management Information Systems: Managing the Digital Firm*, 8th ed., Upper Saddle River, NJ: Prentice Hall, 2004.

McFarlan, F.W. and McKenney, J.L., *Corporate Information Systems Management. The Issues Facing Senior Executives*, Homewood, IL: Dow Jones-Irwin, 1983.

McNurlin, B.C. and Sprague, R.H., *Information Systems Management In Practice*, 6th ed., Upper Saddle River, NJ: Pearson/Prentice Hall, 2004.

Parker, M.M., Benson, R.J. and Trainor, H.E., *Information Economics. Linking Business Performance to Information Technology*, Upper Saddle River, NJ: Prentice Hall, 1988.

Powell, P., "Information Technology Evaluation, is it Different?," *Journal of the Operational Research Society*, 1 (1992), pp. 29-42.

Rainer, R.K., Snyder, C.A. and Carr, H.H., "Risk Analysis for Information Technology," *Journal of Management Information Systems*, Vol. 8, No. 1, 1991, pp. 121-120.

The Costs of Not Making the Right Information Technology Decision (and Strategies on How to Avoid Them)

We can explore the costs of not making the right information technology (IT) investment decision in both a macro economic and micro economic context.

"In a moment of decision, the best thing you can do is the right thing to do. The worst thing you can do is nothing." Theodore Roosevelt, President of the United States

Macro Economic View

In a macro economic sense, a firm operates within an industry and within global markets. The integration of IT creates linkages between all possible stakeholders that are connected to or do business with any firm. *Stakeholders* in a macro economic sense include the external partnering companies that help the firm perform their business functions, their supply-chains that link firms together in their industry and with other supply-chains, the government, and society as whole. If a firm in a particular industry does not make the right IT investment decision, then all stakeholders can be negatively impacted. Examples are presented in Table 1 (McNurlin and Sprague, 2004, pp. 5-12; Turban *et al.,* 2003, pp. 489-511; Wen and Yen, 1998).

"Vacillating people seldom succeed. Successful men and women are very careful in reaching their decisions, and very persistent and determined in action thereafter." L. G. Elliott, writer.

Table 1. Examples of stakeholder costs.

Stakeholder	Examples of macro economic costs
Industry customers	Poor IT decisions could cause the firm to go out of business. That reduces competition and increases the likelihood of higher prices to customers. It also diminishes the quality of selection of products within the industry to all customers.
Industry members	Poor IT decisions could cause the firm to go out of business, which could diminish the industry's demand for supplies, and in turn possibly diminish the supply-chain network that support other companies in the same industry.
Industry partners and supply-chain members	Poor decisions on IT can burden suppliers, vendors, and consultants forcing them to incur needless costs to maintain equally poor IT that may not serve their internal needs. Poor IT and the interfaces across supply-chains can slow down communications, making them less efficient and more costly for all members. Since in economic theory all supply-chains are linked together, diminishing one supply-chain will have a negative impact on all those other supply-chains that are linked to it.
Government	Poor IT can inhibit information between the firm and the government agencies in monitoring problems. Earlier detection of problems and notification by the government might save the firm unnecessary rework costs. Poor IT can also burden the government in their efforts to do a better service for all society and increases the government's costs.
Society	Poor IT can delay, delete, and cancel customer orders causing frustration and costs of all kind. Poor IT investments will eventually be passed on to the consumer, which means needless higher costs to them. Some bad decisions can cause an entire company to go out of business, resulting in the loss of jobs to the employees and revenue to their local economies.

While the examples in Table 1 may be considered esoteric to most IT managers facing an investment decision, it is hoped that they view these macro economic considerations in the same light as they would "ethical factors" in their decision-making. Indeed, *ethics* has been defined as the principles of right and wrong that individuals use to make choices to guide their behavior (Laudon and Laudon, 2004, p. 146). Like a pebble dropped in a lake of water that creates a ripple effect, so to is the magnitude of the impact of a wrong IT decision and its costs to companies throughout a company's supply chain. The fact is bad or wrong IT decisions do cost customers, business partners, and all of society unnecessary resources. To be a responsible IT manager, the right IT investment decisions should always be sought for the greater good of all stakeholders.

"It is change, continuing change, inevitable change, that is the dominant factor in society today. No sensible decision can be made any longer without taking into account not only the world as it is, but the world as it will be.... This, in turn, means that our statesmen, our businessmen, our everyman must take on a science fictional way of thinking." Isaac Asimov, writer.

Micro Economic View

In an effort to show where the information technology (IT) investment decision-making methodologies are used in the micro economic environment of the firm, we introduced in Chapter 1 the multi-step procedure for MIS hierarchical planning of IT systems. As shown in Figure 1, this analysis process requires a number of intensive steps, which can be very involved. The over view in Figure 1 presents a broad-based approach to decision-making that encompasses many elements of corporation-wide decision-making.

"Successful leaders have the courage to take action while others hesitate." John C. Maxwell, writer.

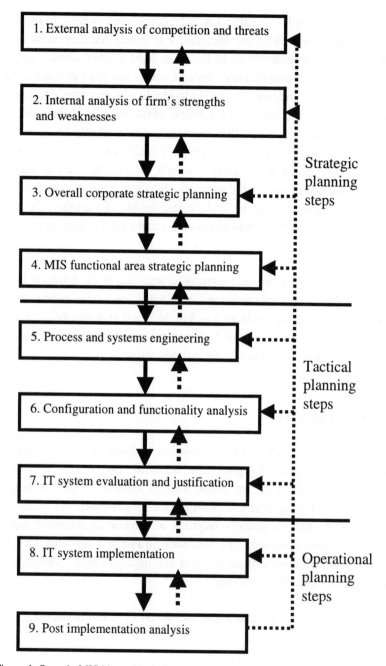

Figure 1. Steps in MIS hierarchical planning of IT systems.

Focusing more precisely on where the IT decision methodologies are applied and how they play a critical role in the investment analysis, we further detailed the tactical steps (note Figure 2) of MIS hierarchical planning from Chapter 2. It is in Step 5 in the tactical planning process where the methodologies presented in Chapters 3 through 12 are applied.

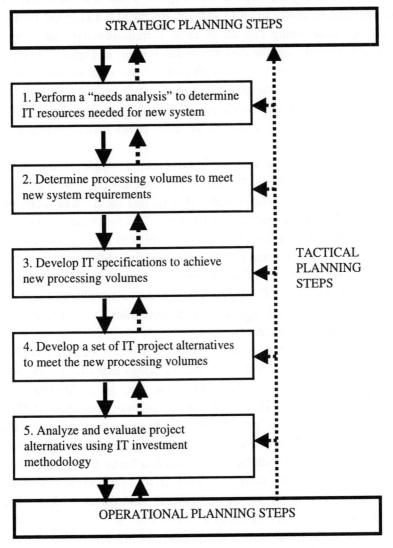

Figure 2. Detailed steps of tactical MIS planning process for IT projects.

"Do the right thing. It will gratify some people and astonish the rest." Mark Twain, writer.

We have list these detailed steps in this epilogue chapter not only as a summary of how this textbook's content is related, but also to point out that anywhere along the steps in both Figures 1 and 2 mistakes can be made that can be costly to an IT manger, department and to the firm as whole. Even assuming that all of these prerequisite steps have been performed correctly and the decision maker is ready in Figure 2's Step 5 to analyze and evaluate IT project alternatives, the right decision on an IT investment may not be possible, unless a full understanding of the costs involved have been considered. This leads us to the subject of this epilogue chapter: How to determine the costs of not making the right IT decision.

IT costs are relative to the objective being sought at the micro economic level of the firm. To understand what any decision is costing, one needs to know the relative *value* of what one seeks. Kauffman and Weill (1989) have suggested that the right IT investment is one that maximizes the value of the firm. They feel this can be accomplished by selecting IT investments that maximize IT benefits while minimizing "IT risks". *IT risks* refer to *IT asset risks, IT staff risks, IT design and development risks*, and *IT implementation risks*. Examples of these are presented in Table 2 and represent the opportunities for increased costs in present or future investments.

Given these costs, we can now discuss the idea of "value". Many IT researchers feel that the *productivity* that IT brings to the firm is the best measure of IT value (Lee, 2001; Shao and Lin, 2002; Krishnan and Sriram, 2000; Hu and Plant, 2001). Unfortunately, the value of IT is often a debated issue as pointed out in Chapter 1's reference to the *productivity paradox* and the many other references cited throughout this textbook. Indeed, research by Chircu and Kauffman (2000) and Chan (2000) have shown that there are limits as to how accurate valuation models can be in determining the real value IT contributes to a firm. The costs and difficulties in estimating IT value leads to a logical question: How do we avoid these costs and accurately estimate IT value to make the right IT investment decisions?

Table 2. Examples of IT risks.

IT asset risks to hardware, software, and data	IT staff risks	IT design and development risks	IT implementation risks
Vulnerability due to access can cause law suits due to revealing sensitive customer information	Employee training exceeds estimates requiring additional expenses	Failure to obtain anticipated benefits adds to cost of operations	Costs that exceed estimates
Vulnerability due to piracy or theft requires replacement costs	Changes in salary to match new technology skills increases expenses	IT unable to support current business operations requires ongoing and additional future expenditures to fix	Time exceeding estimates can cause lost customers and penalty costs
Vulnerability due to purposeful or accidental deletion	Management time exceeds expectations	IT unable to support future business operations requires ongoing and additional future expenditures to fix	Unexpected user resistance to using IT can cause lost productivity adding to operating expenses
Vulnerability due to natural disasters can cause a loss of customers	Employee motivation drops as time increases, requiring overtime expense	Incompatibility or integration system failures requires ongoing and additional future expenditures to fix	Changes cause temporary loss of productivity adding to operating expenses
Vulnerability due to obsolescence can cause a loss of customers and reduce productivity which increases costs	Employee turnover increases due to changes requiring increased costs in training	System design short-lived, requiring replacement well ahead of life expectancy.	IT staff lacks skills and/or unable to successfully run technology

Strategies for Making the Right IT Investment Decisions and Avoiding IT Costs

"It's less important to have unanimity than it is to be making the right decisions and doing the right thing, even though at the outset it may seem lonesome." Donald Rumsfeld, US Secretary of Defense.

Many strategies have been suggested based on research and experience of management information systems experts. Here is a collection of some of the strategies that have proven themselves in IT investment decision-making (Beach *et al.*, 2000; Chan, 2000; Chircu and Kauffman, 2000; Dehning and Richardson, 2002; Hu and Plant, 2001;Irani and Love, 2002; Ross and Wiell, 2002; Shao and Lin, 2002; Li and Ye, 1999):

- *IT value and implementation must be discussed in the context of the organization's goals, strategies, tactics, operational plans, and culture.* In order to determine a payback, you must determine the benefits as they help an organization achieve their goals as pointed out in Chapters 1 and 2. This will also require that stakeholder impacts be determined and monitored, so the relevance of the contribution that IT makes to vested and interest controlling members of the organization is clearly documented and available for review.
- *Executive managers, not IT managers, should determine strategic allocation decisions.* The total amount of funds to invest in IT, which business processes should receive funding and which IT capabilities are needed organization-wide are decisions that executive managers or vice presidents (VPs) should make, not IT managers. This guideline ties back to the strategic planning steps in Chapter 1 where initial organization-wide planning must be done at the executive manager or VP levels of the organization so their plans reflect organization goals and can be filtered down for eventual implementation through the organization. In most organizations only these executives are privy to all the types of sensitivity external and internal information (note Figure 1) necessary for the right kinds of corporation planning to meet competitive challenges external to

the firm. Since a VP of information systems or a Chief Information Officer (CIO) is usually included in this group of executives, input from the IT area will be shared. It is the job of the IT managers, once the vision or strategic plans are laid, to then develop tactics and operational plans to implement the strategic plans.

- *In order to measure IT value and its performance over time, utilize many measures of contribution and performance.* The fastest way to fail in IT value measurement is to limit an analysis of cost or benefits to just a few points in an information system or just a few measures of performance. Information systems and their use of IT are dynamic and integrated into a variety of systems (i.e., marketing, accounting, finance, purchasing, production, etc.). Business performance measures for all areas in firm should be used as a part of the IT valuation process. Another aspect of this guideline is that no single measure should be the major determiner of success or failure. You should always use a combination of measures to make that type of determination. In Chapters 2 through 12 we have presented many methodologies, ratios, and other techniques that can be used to value and measure IT performance. You are encouraged to use as many of these measure as you can reasonably be applied. If there is one thing the predominance of Cost/Benefit Analysis (Chapter 6) has made clear, is that it takes more than one measure to capture the essence of business success or failure. While it may seem excessive at first to do all that analysis, you may find as time goes on that one measure that was not important in the beginning may later become important. In doing so you will find that there is safety in numbers (no joke intended).

- *General guidelines on issues of security and privacy risks, project failure risks, and the quality of IT services should be determined by executive managers and not IT managers.* Security, privacy and project failure risks can be very large risks, involving the potential destruction of the entire organization. They are, therefore, serious enough for executive managers, CIOs or VPs to have a hand in establishing their willingness to access these risks. The executive definition of risk acceptance will provide a very beneficial guideline to help determine the scope of funding possible and measure risk avoidance behavior in valuing IT investments during their use. It is equally important to "pin down" the executive managers on what they consider is quality IT services. Some may define it in terms of speed of service and others in quantity of service. Regardless, without a guideline, spending is guided by either guesses or disasters, and neither represents good planning.

- *IT evaluation methods must evolve with the organization. Organizations change over time, IT adapts to those changes, and the measurement methods and systems used to monitor and assess the value of IT must also change.* We need to move away from the logic of IT investment being a short-term event and view it as a long-term relationship with multiple events shaping its real value to the organization. By using a wide variety of IT investment methodologies, unique capabilities representing the current and future value of IT may be captured more purposefully for valuation purposes.

- *Recognize the limitations of the IT investment methodologies at each phase of the IT investment decision process.* As we have repeatedly mentioned in all the chapters of this textbook that all of the IT investment methodologies presented herein have limitations. These limitations may in some cases disqualify methodologies from being applied, and rightfully so. The assumptions under which models like Goal Programming in Chapter 9 and Game Theory in Chapter 10 are very restrictive, but they have to be in order to be assured that near-optimal solutions are valid in application. Other less restrictive models, like Accounting Rate of Return in Chapter 4 and Present Value Analysis in Chapter 5 are less versatile in application than those methods requiring more rigorous qualifications. Fundamental to all methodology use is the need to be aware and accept those limitations in the output of the specific IT investment methodologies used in an analysis.

- *Recognize in all the selection processes mentioned above that the IT manager has potential biases that can preclude the right IT decision choice from the analysis.* In this textbook, we have seen repeatedly how it takes both objective and subjective criteria to render the right IT investment decision. Indeed, the identification of relevant criteria is critical in the selection of the

IT investment methodology to use to make the IT investment decision. Making the right decision on both the criteria to include in the IT investment decision analysis and the methodology to use requires decision-making skills that make us aware of factors that may bias our decision process. A listing of these roadblocks to the decision-making process are presented in Table 3. Our human nature and our education can make us act in biased ways, regardless of our good intentions. No one can eliminate these biases completely since we are human and we are educated to do things in particular ways. What we can do is minimize their influence by being aware that each can negatively impact our decision process, and then make extra effort to double check our final decisions to see if we have in fact been bias in some way. In other words, we need to exercise self-awareness in our decision-making efforts. In doing so, we can compensate for the biases and adjust our decisions accordingly.

"It doesn't matter which side of the fence you get off on sometimes. What matters most is getting off. You cannot make progress without making decisions." Jim Rohn, business executive and best-selling author.

"So What's the Good News?"

We can not end this textbook without reiterating why it's content and the work that is required to use it is worth the effort. What is the "good news" for the IT manager when making the right decisions? If you make the right IT decisions, you will end up with an investment that helps your organization to introduce, create or enhance a competitive advantage. Let's look at a small set of examples reported in the literature:

- *Improving organization agility*: One of the most important competitive advantages in today's markets is the ability to be agile or develop the capacity to react quickly and successfully to change so as to compete effectively in many developed and emerging global markets. McGaughey (1999) found that investments in Internet technology, in the form of corporate intranets and the Internet, was an important enabler of agility. These IT investments help make possible the intra- and interorganizational sharing of data and information in the form of text, graphics, audio, and video, enabling various tasks, activities, and processes that help a firm to become agile and better able to compete.

Table 3. Possible bias in our decision process.

Decision making biases	Description and resolution
Direct decision toward single goal	The tendency to view all problems in light of a single goal (i.e., profit maximization). Be opened minded and realize that a problem can be viewed from many perspectives at the same time (i.e., profit, quality, cost, etc.).
Equate new with old experiences	The tendency to try and solve new problems with old solutions. Treat each new problem uniquely as it deserves.
Use available solutions	The tendency for IT people to look at all problems in terms of just an IT solution. If you have a human resource problem in IT, it's a human resource problem and should be dealt with as such.
Confuse symptoms with problems	The tendency to try and solve a symptom, rather than get at the root problem. If a computer stops printing, you should fix the technology, not just buy more paper in hopes it will start printing again.
Deal with problems at face value	The tendency to jump at a solution before the problem has been investigated. IT problems can be very complex and require a thorough investigation of cause and effect factors that contribute to its existence before a real solution should be considered.
Using value judgments	The tendency to base a decision on a referent. If we are told a particular manufacturer's technology is of high quality, we might start using that manufacturer as a "benchmark" without having investigated or fairly evaluated them ourselves. We need to be assured that our referents or benchmarks are what we are using them to be.
To discount possible solutions	The tendency to overlook possible solutions to long-term problems because you feel they can not be solved. In the long-run, nothing is unsolvable. Opportunities to solve endemic or perpetual problems can surface anytime, particular with the advances in IT. Don't close your mind to new ideas that can solve old problems.

- *Helps organizations adjust marketing mix factors to better compete*: As markets for products change, so do what customers look for in a product change. Successful firms must change their marketing efforts to match the consumer expectations in the mix features offered with a product. Bramorski, *et al.* (2000) found that price had traditionally been the order-winning criterion for ready-to-assemble (RTA) products such as furniture and bicycles. Changes in today's market increasingly emphasize quality, not price, as the order-winning criterion, along with time-based speed-of-delivery and flexibility in customer ordering. This research reviewed how IT investments permit organizations to change their RTA products and processes to effectively improve information to customers and help the products deliver higher quality and better service.

- *A means of identifying strategic external competitive intelligence*: Implicitly or explicitly an organization's strategic intelligence is a pre-requisite for change, and that effective investments in IT represent a critical requirement for implementing the changes that will take place as a results of that intelligence. Guimaraes (2000) described how organizations are identifying strategic problems and opportunities and how to effectively implement business changes by using IT. The research revealed empirical evidence about the importance of competitive intelligence and IT support for effectively implementing change in business organizations. The research also showed how managers should go about acquiring competitive intelligence and managing IT to effectively support business improvements.

- *Reducing IT investment costs*: The right decision on IT investment sometimes means not making an investment as all. As Chabrow (2003) has observed in recent IT investments, many firms today are now using outsourcing, leasing, renting or obtaining the technology they need by shifting its cost to an outside maintenance organization that pays for the technology they maintain for client firms. Unless the various financial and non-financial analyses mentioned in this textbook are undertaken, it is impossible to know for sure if this popular approach to IT investment decision-making is a worthwhile strategy to follow.

- *Reducing IT operating costs:* According to industry analysts, an enterprise employing 1,000 knowledgeable workers wastes $48,000 per week, or nearly $2.5 million per year, due to an inability to locate and retrieve information (Madjiah, 2003). Toyota Motor America estimates it will spend $120 per computer workstation annually over the next two years on anti-spam efforts, and Siemens VDO Automotive says it spends about $5 per e-mail mailbox a year to filter spam (Kisiel *et al.*, 2003). In the summer of 2001, the Code Red worm infected hundreds of thousands of computers around the world. While this computer virus was fairly benign it nevertheless did a great deal of damage. Many companies lost the use of their networks and their Web sites. The total bill for the mess has been estimated at $2.6 billion. Also, according to industry estimates, security breaches impact 90 percent of all businesses every year and cost some $17 billion. Protective measures are expensive, where an average company can easily spend 5 to 10 percent of its IT budget on security (Austin and Darby, 2003). These operating cost situations all represent opportunities for IT managers to do a better job by making the right decision on IT investments. Research shows, that IT managers can reduce these catastrophes while still cutting costs if they make the right IT investments ("Computer Virus Costs", 2002; Erhun and Tayur, 2003).

"One thing is sure. We have to do something. We have to do the best we know how at the moment...; If it doesn't turn out right, we can modify it as we go along." Franklin D. Roosevelt, President of the United States

References

Austin, R.D. and Darby, C.A.R., "The Myth of Secure Computing," *Harvard Business Review*, Vol. 81, No. 6, 2003, pp. 120-127.

Beach, R., Muhlemann, A.P., Price, D.H., Paterson, A. and Sharp, J.A., "The Selection of Information Systems for Production Management: An

Evolving Problem," *International Journal of Production Economics*, Vol. 64, 2000, pp. 319-329.

Bramorski, T., Madan, M., Motwani, J. and Sundarraj, R.P., "Improving Competitiveness of Ready-to-assemble Manufacturers Through Information Technology," *Logistics Information Management*, Vol. 13, No. 4, 2000, pp. 201-209.

Chabrow, E., "Creative Pressure," *Information Week*, No. 944, June 2003, pp. 38-44.

Chan, Y.E., "IT Value: The Great Divide Between Qualitative and Quantitative and Individual and Organizational Measures," *Journal of Management Information Systems*, Vol. 16, No. 4, 2000, pp. 225-261.

Chircu, A.M. and Kauffman, R.J., "Limits to Value in Electronic Commerce-Related IT Investments," *Journal of Management Information Systems*, Vol. 17, No. 2, 2000, pp. 59-80.

"Computer Virus Costs," *Controller's Report*, Vol. 2002, No. 8, (August 2002), p. 15.

Dehning, B. and Richardson, V.J., "Returns on Investments in Information Technology: A Research Synthesis," *Journal of Information Systems*, Vol. 16, No. 1, 2002, pp. 7-30.

Erhun, F. and Tayur, S., "Enterprise-wide Optimization of Total Landed Cost at a Grocery Retailer," *Operations Research*, Vol. 51, No. 3, 2003, pp. 343-354.

Guimaraes, T., "The Impact of Competitive Intelligence and IS Support in Changing Small Business Organizations," *Logistics Information Management*, Vol. 13, No. 3, 2000, pp. 117-125.

Hu, Q. and Plant, R., "An Empirical Study of the Casual Relationship Between IT Investment and Firm Performance," *Information Resources Management Journal*, Vol. 14, No. 3, 2001, pp. 15-26.

Irani, Z. and Love, P., "Developing a Frame of Reference for Ex-ante IT/IS Investment Evalution," *European Journal of Information Systems*, Vol. 11, 2002, pp. 74-82.

Kisiel, R., Moran, T. and Osburn, C., "Battle Against Spam," *Automotive News*, Vol. 77 No. 6046, 2003, pp. 1IT-3IT.

Krishnan, G.V. and Sriram, R.S., "An Examination of the Effect of IT Investments on Firm Value: The Case of Y2K-Compliance Costs," *Journal of Information Systems*, Vol. 14, No. 2, 2000, pp. 95-108.

Laudon, K.C. and Laudon, J.P., *Management Information Systems: Managing the Digital Firm*, 8th ed., Upper Saddle River, NJ: Prentice Hall, 2004.

Lee, C.S., "Modeling the Business Value of Information Technology," *Information & Management*, Vol. 39, 2001, pp. 191-210.

Li, M. and Ye, R., "Information Technology and Firm Performance: Linking with Environmental, Strategic and Managerial Contexts," *Information & Management*, Vol. 35, 1999, pp. 43-51.

Madjiah, L.E., "Oracle Collaboration Suite: All in One Inbox ," *Jakarta Post*, July 5, 2003 (INDONESIA).

McGaughey, R.E., "Internet Technology: Contributing to Agility in the Twenty-First Century," *International Journal of Agile Management Systems*, Vol. 1, No. 1, 1999, pp. 7-13.

McNurlin, B.C. and Sprague, R.H., *Information Systems Management In Practice*, 6th ed., Upper Saddle River, NJ: Pearson/Prentice Hall, 2004.

Ross, J.W. and Weill, P., "Six IT Decisions Your IT People Shouldn't Make," *Harvard Business Review*, November, 2002, pp. 84-91.

Shao, B. and Lin, W.T., "Technical Efficiency Analysis of Information Technology Investments: A Two-Stage Empirical Investigation," *Information & Management*, Vol. 39, 2002, pp. 391-401.

Turban, E., Rainer, R.K. and Potter, R.E., *Introduction to Information Technology*, 2nd ed., New York, NY: John Wiley & Sons, Inc., 2003.

Wen, J.H. and Yen, D.D., "Methods for Measuring Information Technology Investment Payoff", *Human Systems Management*, Vol. 17, No. 2, (1998), pp. 145-153.

Index